Reproduction in mammals
Book 3: Hormonal control of reproduction

SECOND EDITION **Reproduction in mammals**

BOOK **3** *Hormonal control of reproduction*

EDITED BY C. R. AUSTIN

Formerly Fellow of Fitzwilliam College
Emeritus Charles Darwin Professor of Animal Embryology
University of Cambridge

AND R. V. SHORT, FRS

Professor of Reproductive Biology
Monash University, Melbourne, Australia

DRAWINGS BY JOHN R. FULLER

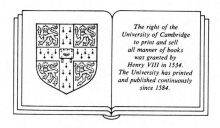

The right of the
University of Cambridge
to print and sell
all manner of books
was granted by
Henry VIII in 1534.
The University has printed
and published continuously
since 1584.

Cambridge University Press
Cambridge
New York   New Rochelle
Melbourne   Sydney

Published by the Press Syndicate of the University of Cambridge
The Pitt Building, Trumpington Street, Cambridge CB2 1RP
32 East 57th Street, New York, NY 10022, USA
10 Stamford Road, Oakleigh, Melbourne 3166, Australia

First published 1972
Reprinted 1973, 1975, 1978
Second edition 1984
Reprinted 1986, 1987, 1988

Printed in Great Britain by the University Press, Cambridge

Library of Congress catalogue card number: 83-7506

British Library Cataloguing in Publication Data
Austin, C. R.
Reproduction in mammals. – 2nd ed.
Book 3: Hormonal control of reproduction
1. Mammals – Reproduction
I. Title II. Short, R. V.
599.01′6 QL739.2

ISBN 0 521 25637 2 hard covers
ISBN 0 521 27594 6 paperback
(First edition:
ISBN 0 521 08438 5 hard covers
ISBN 0 521 09696 0 paperback)

CONTENTS

Contents

CONTRIBUTORS TO BOOK 3

D. T. Baird
Department of Obstetrics and Gynaecology
37 Chalmers Street
Edinburgh EH3 9EW, UK

A. T. Cowie
6 Maiden Erlegh Drive
Earley Drive
Reading RG6 2HP, UK

A. P. F. Flint
ARC Institute of Animal Physiology
Babraham
Cambridge CB2 4AT, UK

R. B. Heap
ARC Institute of Animal Physiology
Babraham
Cambridge CB2 4AT, UK

F. J. Karsch
Department of Physiology and Center for Human Growth
and Development
The University of Michigan
Ann Arbor
Michigan 48109, USA

D. M. de Kretser
Department of Anatomy
Monash University
Clayton
Victoria 3168, Australia

D. W. Lincoln
Centre for Reproductive Biology
37 Chalmers Street
Edinburgh EH3 9EW, UK

G. A. Lincoln
Centre for Reproductive Biology
37 Chalmers Street
Edinburgh EH3 9EW, UK

R. V. Short, FRS
Department of Physiology
Monash University
Clayton
Victoria 3168, Australia

PREFACE TO THE SECOND EDITION

In this, our Second Edition of *Reproduction in Mammals*, we are responding to numerous requests for a more up-to-date and rather more detailed treatment of the subject. The First Edition was accorded an excellent reception, but the Books 1 to 5 were written ten years ago and inevitably there have been advances on many fronts since then. As before, the manner of presentation is intended to make the subject matter interesting to read and readily comprehensible to undergraduates in the biological sciences, and yet with sufficient depth to provide a valued source of information to graduates engaged in both teaching and research. Our authors have been selected from among the best known in their respective fields.

Book 3 discusses the manifold ways in which hormones control the reproductive processes in male and female mammals. The hypothalamus regulates both the anterior and posterior pituitary glands, whilst the pineal can exert a modulating influence on the hypothalamus. The pituitary gonadotrophins regulate the endocrine and gametogenic activities of the gonads, and there are important local feedback effects of hormones within the gonads themselves. Non-pregnant females display many different types of oestrous or menstrual cycles, and there are likewise great species differences in the endocrinology of pregnancy. But the hallmark of mammals is lactation, and this also exerts a major control on subsequent reproductive activity.

From the Preface to the First Edition
Reproduction in Mammals is intended to meet the needs of undergraduates reading Zoology, Biology, Physiology, Medicine, Veterinary Science and Agriculture, and as a source of information for advanced students and research workers. It is published as a series of eight small textbooks dealing with all major aspects of mammalian reproduction. Each of the component books is designed to cover independently fairly distinct subdivisions of the subject, so that readers can select texts relevant to their particular interests and needs, if reluctant to purchase the whole work. The contents lists of all the books are set out on the next page.

BOOKS IN THE FIRST EDITION

Book 1. Germ cells and fertilization
Primordial germ cells *T. G. Baker*
Oogenesis and ovulation *T. G. Baker*
Spermatogenesis and the spermatozoa *V. Monesi*
Cycles and seasons *R. M. F. S. Sadleir*
Fertilization *C. R. Austin*
Book 2. Embryonic and fetal development
The embryo *A. McLaren*
Sex determination and differentiation *R. V. Short*
The fetus and birth *G. C. Liggins*
Manipulation of development *R. L. Gardner*
Pregnancy losses and birth defects *C. R. Austin*
Book 3. Hormones in reproduction
Reproductive hormones *D. T. Baird*
The hypothalamus *B. A. Cross*
Role of hormones in sex cycles *R. V. Short*
Role of hormones in pregnancy *R. B. Heap*
Lactation and its hormonal control *A. T. Cowie*
Book 4. Reproductive patterns
Species differences *R. V. Short*
Behavioural patterns *J. Herbert*
Environmental effects *R. M. F. S. Sadleir*
Immunological influences *R. G. Edwards*
Aging and reproduction *C. E. Adams*
Book 5. Artificial control of reproduction
Increasing reproductive potential in farm animals *C. Polge*
Limiting human reproductive potential *D. M. Potts*
Chemical methods of male contraception *H. Jackson*
Control of human development *R. G. Edwards*
Reproduction and human society *R. V. Short*
The ethics of manipulating reproduction in man *C. R. Austin*
Book 6. The evolution of reproduction
The development of sexual reproduction *S. Ohno*
Evolution of viviparity in mammals *G. B. Sharman*
Selection for reproductive success *P. A. Jewell*
The origin of species *R. V. Short*
Specialization of gametes *C. R. Austin*
Book 7. Mechanisms of hormone action
Releasing hormones *H. M. Fraser*
Pituitary and placental hormones *J. Dorrington*
Prostaglandins *J. R. G. Challis*
The androgens *W. I. P. Mainwaring*
The oestrogens *E. V. Jensen*
Progesterone *R. B. Heap and A. P. F. Flint*
Book 8. Human sexuality
The origins of human sexuality *R. V. Short*
Human sexual behaviour *J. Bancroft*
Variant forms of human sexual behaviour *R. Green*
Patterns of sexual behaviour in contemporary society *M. Schofield*
Constraints on sexual behaviour *C. R. Austin*
A perennial morality *G. R. Dunstan*

x

BOOKS IN THE SECOND EDITION

Book 1. Germ cells and fertilization
Primordial germ cells and the regulation of meiosis *A. G. Byskov*
Oogenesis and ovulation *T. G. Baker*
The egg *C. R. Austin*
Spermatogenesis and spermatozoa *B. P. Setchell*
Sperm and egg transport *M. J. K. Harper*
Fertilization *J. M. Bedford*

Book 2. Embryonic and fetal development
The embryo *A. McLaren*
Implantation and placentation *M. B. Renfree*
Sex determination and differentiation *R. V. Short*
The fetus and birth *G. C. Liggins*
Pregnancy losses and birth defects *P. A. Jacobs*
Manipulation of development *R. L. Gardner*

Book 3. Hormonal control of reproduction
The hypothalamus and anterior pituitary gland *F. J. Karsch*
The posterior pituitary *D. W. Lincoln*
The pineal gland *G. A. Lincoln*
The testis *D. M. de Kretser*
The ovary *D. T. Baird*
Oestrous and menstrual cycles *R. V. Short*
Pregnancy *R. B. Heap and A. P. F. Flint*
Lactation *A. T. Cowie*

Book 4. Reproductive fitness
Reproductive strategies *R. M. May and D. Rubenstein*
Species differences in reproductive mechanisms *R. V. Short*
Genetic control of fertility *R. B. Land*
Effects of the environment on reproduction *B. K. Follett*
Sexual and maternal behaviour *E. B. Keverne*
Immunological factors in reproductive fitness *N. J. Alexander and D. J. Anderson*
Reproductive senescence *C. E. Adams*

Book 5. Manipulating reproduction
Increasing productivity in farm animals *K. J. Betteridge*
Today's and tomorrow's contraceptives *R. V. Short*
Contraceptive needs of the developing world *D. M. Potts*
Risks and benefits of contraception *M. P. Vessey*
Improving human fertility *J. Cohen, C. B. Fehilly and R. G. Edwards*
Our reproductive options *A. McLaren*
Barriers to population control *C. R. Austin*

1

The hypothalamus and anterior pituitary gland

FRED J. KARSCH

It should not come as a surprise that the reproductive process in mammals is governed by the central nervous system. Indeed, people have recognized since biblical times that breeding activity in many species is confined to particular seasons of the year, thus pointing to a role for the nervous system

Fig. 1.1. Organization of control systems governing reproduction.

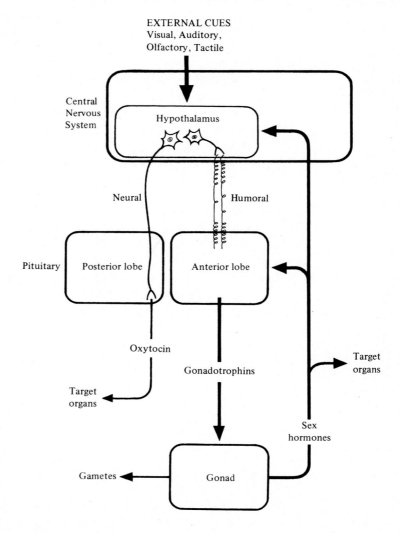

in perceiving information about the environment and conveying this information to the gonads. What is surprising is that the specific mechanisms by which the nervous system exercises this control have, until recently, remained a mystery. Not until the discovery by Geoffrey Harris and John Green in the 1940s, that the brain can direct the activity of the anterior lobe of the pituitary gland, did the details of this control system begin to be fully appreciated. Even to this day, many important aspects of this regulation have not been defined.

An organizational overview of the system that governs reproduction is provided in Fig 1.1. It is now clear that information emanating from a variety of external cues (eg. visual, auditory, tactile, olfactory) is fed into the central nervous system and converges on the hypothalamus. There, the information is processed, amplified, transduced to a humoral signal, and transmitted to the anterior pituitary gland where it is further amplified and transmitted, via the gonadotrophic hormones, to the gonads. The latter respond in many ways, one of which is by the secretion of sex hormones. These, in turn, act on a host of target tissues including the brain and pituitary gland. This forms a vastly complex network of information transfer, a network that permits amplification, propagation, and integration of signals throughout the body.

In this chapter, we will focus on the interrelationship between the brain and the anterior lobe of the pituitary gland. Additional coverage of this topic is provided in Chapter 1, Book 7, of the First Edition of this series. The remaining chapters of the present book are directed towards other aspects of this control system.

Anatomical organization of the hypothalamo-pituitary axis
Hypothalamus
The brain consists of lobular outgrowths from the walls of the fluid-filled neural tube. The central cavities of these outgrowths are called ventricles. They contain the cerebrospinal fluid, a filtrate of blood into which a variety of substances produced in the brain are secreted.

The hypothalamus forms the base of the brain in the region of the diencephalon, a subdivision of the forebrain (Fig. 1.2). The hypothalamic boundary is limited frontally by the optic chiasma, caudally by the mamillary bodies, and dorsally by the thalamus (another region of the diencephalon). The hypothalamus surrounds the fluid-filled third ventricle. It is made up of various types of structural elements including cell bodies of hypothalamic neurones with their axons and terminals, axons and terminals of other neurones with cell bodies lying outside the hypothalamus, and axons passing through from extrahypothalamic neurones. It also contains glial cells which form the supporting structure for neurones. Finally, it has a very special type of blood supply, particularly in the region of the median eminence (see below).

The cell bodies of hypothalamic neurones are not scattered diffusely throughout the glial cells, but are clustered in discrete regions called nuclei. There are many such nuclei, only some of which are important for the control of reproduction. The cell bodies of the hypothalamic nuclei send axonal projections to one of four general regions: (1) other areas of the brain, (2) other hypothalamic nuclei, (3) the median eminence, and (4) the posterior lobe of the pituitary gland (Fig. 1.3). This arrangement provides for an extremely complex network of neuronal communication and interaction among both neural and endocrine regulatory centres.

The median eminence region deserves special mention because it is an area of confluence of neural and blood-borne messages that regulate the function of the anterior lobe of the pituitary gland. The median eminence comprises the base of the hypothalamus and is continuous with the pituitary stalk. It contains few, if any, nerve cell bodies, but consists of axons and terminals of both hypothalamic and extrahypothalamic neurones, glial cells, and specialized ependymal cells called tanacytes. The latter cells line the third ventricle and are of potential importance in the transfer of information from the cerebrospinal fluid to the pituitary gland.

Fig. 1.2. Anatomical organization of the hypothalamo-hypophyseal axis. Arrows designate relative amount of blood flow in each direction in the pituitary stalk. MN, magnocellular endocrine neurone; PN, parvicellular endocrine neurone. (Adapted from F. H. Netter. In *The Ciba Collection of Medical Illustrations*, vol. 4, *Endocrine System and Selected Metabolic Disorders*, CIBA; New York (1965).)

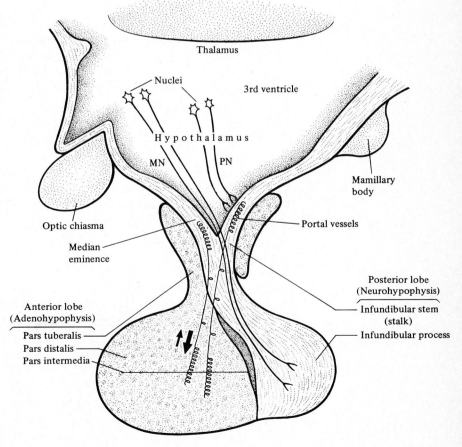

In addition to these neural structures, the median eminence contains a capillary plexus connected with the hypothalamo-pituitary portal system.

Because of this anatomical arrangement and its proximity to the pituitary gland, the hypothalamus assumes special significance as an interface between the central nervous system and the endocrine system. It may be viewed as a vastly complex switchboard connecting the brain above to the endocrine system below.

Pituitary gland

The pituitary gland, or hypophysis, consists of two major subdivisions, the anterior lobe (or adenohypophysis) and the posterior lobe (or neurohypophysis) (Fig. 1.2). The posterior lobe is further divided into the infundibular process and the infundibular stem (or pituitary stalk) which connects to the median eminence above. The posterior lobe is made up of neural tissue and is connected to the rest of the brain via the stalk. The infundibular process contains terminals of neurones whose cell bodies reside in the hypothalamus. Thus, there is a direct neural link between the posterior pituitary and the brain. This is particularly important for the secretion of hormones from the posterior pituitary (see Chapter 2). The anterior lobe of the pituitary (or adenohypophysis) is further subdivided into the pars distalis, pars intermedia, and pars tuberalis. The pars tuberalis surrounds the infundibular stem like a cuff and extends upwards to lie beneath a portion of the median eminence. Unlike the posterior lobe, the anterior pituitary contains no nerve fibres and terminals and so is not in direct neuronal contact with the hypothalamus. Instead, it is connected to the brain by a vascular connection, the hypothalamo-hypophyseal portal system.

Fig. 1.3. Diagram illustrating major projections of hypothalamic neurones. Numbers indicate the four different types of hypothalamic nuclei. AP, anterior pituitary; PP, posterior pituitary; OC, optic chiasma; ME, median eminence; MB, mamillary body.

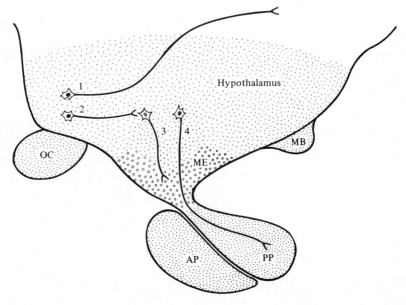

The hypothalamo-hypophyseal portal system. This system has its primary capillary plexus in the median eminence. Its vessels course down the pituitary stalk and terminate in the secondary capillary plexus within the anterior lobe (Fig. 1.2). Most of the blood supplied to the anterior lobe comes from this portal system; in some species, it accounts for over 90 per cent of the total. Thus, most of the blood reaching the anterior lobe travels in vessels first bathed by extracellular fluid in the capillary plexus of the median eminence, a region rich in terminals of the hypothalamic neurones.

The existence of the hypothalamo-hypophyseal portal system was not recognized until the late 1920s and early 1930s. At first, it was thought that blood flowed up the stalk from the pituitary to the hypothalamus. Largely through the efforts of George Wislocki it was eventually established that blood flows down the stalk, thereby providing a route for information transfer from the brain to the anterior pituitary. Recently we have come to know that, in addition, a small proportion of blood may actually flow up the pituitary stalk, thus providing a direct vascular link from the anterior pituitary back to the hypothalamus.

The anterior pituitary gland consists of many different cell types classified on the basis of their size, shape, and histological staining characteristics. With perhaps one exception, there is a separate cell type for the synthesis and secretion of each of the six known anterior pituitary hormones. The exception is the cell type for the gonadotrophic hormones, namely luteinizing hormone (LH) and follicle stimulating hormone (FSH). Most workers agree that LH and FSH are synthesized and secreted by the same cell, referred to as a gonadotroph. There is controversy on this point, however, and there are those who claim that some gonadotroph cells secrete only one hormone. This becomes important when we consider the mechanisms regulating the differential secretion of LH and FSH in certain physiological circumstances (see Chapter 4). The other cell type that is of

Fig. 1.4. Schematic representation of sources and structures of anterior pituitary hormones of major importance to reproduction. Note FSH and LH come from the same cell and have the same α-subunit structure.

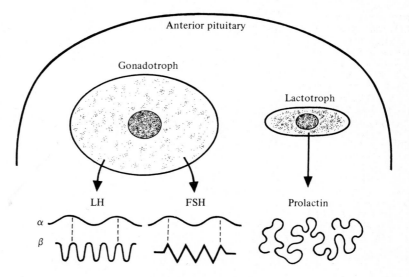

particular significance to reproduction is the lactotroph, the cell that secretes prolactin. The remaining anterior pituitary hormones, like thyroid-stimulating hormone, adrenocorticotrophic hormone and growth hormone, play only a minor role in the regulation of reproduction and thus will not be considered further.

LH and FSH are glycoproteins consisting of two peptide chains, the alpha and beta subunits (Fig. 1.4). The alpha chains of LH and FSH are identical, whereas their beta chains differ and thus confer biological specificity. Both alpha and beta chains are needed for biological activity. Prolactin consists of a single peptide chain. There is considerable species variation in the exact amino acid sequence of the gonadotrophic hormones; there may even be subtle differences in gonadotrophin structure between males and females within a species. Nevertheless, preparations of LH and FSH from one sex are certainly biologically active in the other, whereas this may not always be true of preparations from different species.

Functional organization of the hypothalamo-pituitary axis
Endocrine neurones
Most neurones discharge neurotransmitter substances from their terminals at a synapse; these substances then modulate the activity of other post-synaptic neurones. Many hypothalamic neurones are somewhat different in that they do not release neurotransmitters from their terminals at a synapse; instead they release hormones into the blood stream. These neurones are called endocrine neurones. The hormones they release then modulate the function of cells at a distant site. In most other ways, endocrine neurones are similar to conventional neurones: they have cell bodies, dendrites, and axons with terminal dilatations; they are capable of generating action potentials; and their activity is influenced by traditional neurones that impinge upon them at synapses through the action of neurotransmitters (Fig. 1.5).

There are two general types of endocrine neurones in the hypothalamus: the magnocellular and the parvicellular neurones. As their name implies, the magnocellular neurones are large. Their cell bodies lie in the supraoptic and paraventricular nuclei, and their axons course through the hypo-

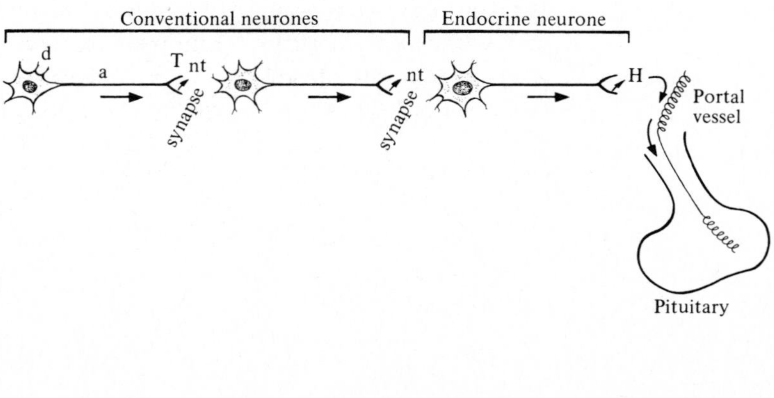

Fig. 1.5. Representation of conventional neurones (left) and endocrine neurones (right) that control reproduction. Note that conventional neurones release neurotransmitters (nt) into synapses, and endocrine neurones release hormones (H) into portal vessels. d, dendrites; a, axon; T, terminal. Arrows indicate direction of information flow.

thalamus to the median eminence and then down the pituitary stalk to terminals in the posterior pituitary gland (see Fig. 1.2; MN). The magnocellular neurones synthesize and secrete the posterior pituitary hormones, oxytocin and vasopressin. Parvicellular neurones are much smaller; their cell bodies are clustered in many hypothalamic nuclei and their axons usually terminate in the median eminence (see Fig. 1.2; PN).

Owing to their large size, a great deal more is known about the functional activity of the magnocellular neurones and this is described in Chapter 2. Their hormones are produced in the cell bodies, packaged into storage granules which are transported down the axons to the terminals, and released from the terminals upon the arrival of action potentials.

Releasing and inhibiting hormones

How does the hypothalamus direct the activity of the anterior pituitary? The parvicellular endocrine neurones synthesize hormones that stimulate or inhibit release of hormones from the anterior pituitary gland. These releasing–inhibiting hormones are discharged from nerve terminals in the median eminence, where they diffuse into the capillaries of the hypothalamo-hypophyseal portal vessels. From there they are carried in the vascular plexus that surrounds the pituitary stalk and distributed to cells throughout the anterior pituitary gland (Fig. 1.5).

Although the existence of hypothalamic releasing factors was recognized in the 1940s, it was not until about 1970 that the first ones were characterized structurally and synthesized chemically, largely through the independent efforts of two Nobel laureates, Roger Guillemin and Andrew Schally. It was originally thought that a separate hypothalamic hormone existed for each anterior pituitary hormone. Once their structures became known and synthetically pure preparations became available, however, it was found that a great deal of overlap exists. For example, the hormone that stimulates the release of LH also induces the secretion of FSH, and thus it is called gonadotrophin-releasing hormone (GnRH). Some workers still hold to the view that there is also another factor, which selectively elicits the release of FSH. Another example of overlap is provided by thyrotrophin-releasing hormone (TRH) which, in large doses, also causes the secretion of prolactin. We do not yet know, however, whether TRH is involved in the physiological regulation of prolactin secretion.

The chemical identity of these releasing and inhibiting hormones, and their mechanism of action on the cells of the anterior pituitary, are described in Chapter 1 of Book 7 (First Edition). All we need say here is that LH and FSH are under stimulatory control from GnRH, whereas prolactin is predominantly under inhibitory control from prolactin-inhibiting factor (PIF), which most people agree is dopamine. The hypothalamic releasing hormones are all peptides. As yet, no hypothalamic hormone has been identified that acts on the pituitary to inhibit LH or FSH secretion.

Transport of releasing–inhibiting hormones

How are hypothalamic hormones carried from their sites of synthesis in neurones to the portal system in the median eminence? Progress in this area has been hampered considerably by difficulty in pin-pointing the specific hypothalamic nuclei that synthesize the releasing–inhibiting hormones. This difficulty stems, in part, from the likelihood that these hormones do not accumulate at their sites of synthesis in cell bodies, but,

Fig. 1.6. Distribution of GnRH in the hypothalamus. The more densely shaded areas indicate regions where the releasing hormone has been located. POA, preoptic area; ME, median eminence; OC, optic chiasma; MB, mamillary body; AP, anterior pituitary; PP, posterior pituitary.

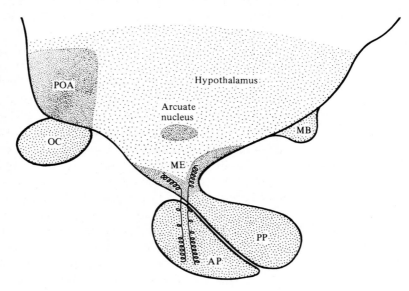

Fig. 1.7. Diagram of proposed routes of transport of releasing hormones (RH) from their sites of production to the portal blood vessels, and the different arrangements (discussed in the text) by which conventional neurones communicate with endocrine neurones. EN, endocrine neurone; nt, neurotransmitter; T, tanacyte; CSF, cerebrospinal fluid.

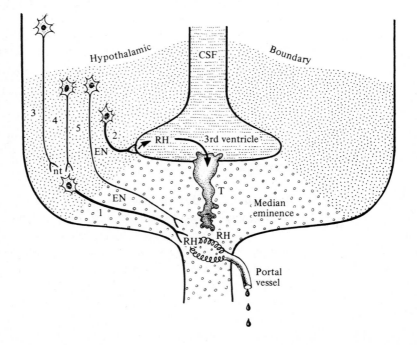

as in the magnocellular system, are transported rapidly down axons to storage depots in nerve terminals. Nevertheless, with the highly sensitive techniques of radioimmunoassay and immunocytochemistry, GnRH has been localized in cell bodies in the medial basal region of the hypothalamus (arcuate nucleus) and in more anterior regions as well (anterior hypothalamus and preoptic area), although there is great species diversity in this regard (Fig. 1.6).

There are two general schools of thought as to how the hypothalamic releasing hormones reach the portal vessels in the median eminence: via axons and via cerebrospinal fluid (Fig. 1.7). According to the axonal transport theory, the endocrine neurones that synthesize the releasing–inhibiting factors have axons that project to terminals abutting on capillaries of the portal vessels in the median eminence. Thus, the hormones are transported to the median eminence via the same cells as those in which they are produced (neurone 1 in Fig. 1.7).

According to the cerebrospinal fluid transport theory, at least some endocrine neurones have axons that project to terminals at the interface between the hypothalamus proper and the fluid-filled third ventricle (neurone 2 in Fig. 1.7). There, the hypothalamic hormones are secreted into the ventricle and transported via the cerebrospinal fluid to the region of the median eminence, where they are picked up by specialized cells called tanacytes. The tanacytes project across the median eminence to terminal-like structures in close proximity to the portal vessels. The tanacytes discharge the releasing hormones, thus bridging the gap between the cerebrospinal fluid and the portal vessels. In this manner, releasing hormones secreted initially into the third ventricle are ultimately carried to the anterior pituitary.

Although the axonal and cerebrospinal fluid transport theories are not mutually exclusive, there is far more evidence in support of the former; the evidence for cerebrospinal fluid transport of hypothalamic hormones is largely anatomical.

Having established the anatomical and functional organization of the system that permits flow of information from the brain to the anterior pituitary, let us now examine how the activity of endocrine neurones is controlled and how this relates to the regulation of gonadotrophin secretion.

Regulation of endocrine neurones

Two general types of control act in concert to regulate the activity of the endocrine neurones relevant to reproduction: control by other neurones through synaptic neurotransmitters, and control by hormones delivered by the blood. Each of these is now considered separately.

Neuronal regulation (*neurotransmitters*)

As mentioned earlier, there are complex neuronal networks linking the hypothalamus with the rest of the brain. These impinge upon the hypothalamic endocrine neurones through a number of different arrangements, some of which are illustrated in Fig. 1.7. These include neurones whose cell bodies reside outside the hypothalamus (neurone 3 in Fig. 1.7) and others whose cell bodies are contained within the hypothalamus proper (neurone 4). Either type may form a synapse with the endocrine neurone in the region of its cell body (neurone 3 and 4), or at its terminal (neurone 5). A single endocrine neurone may be in synaptic contact with many conventional neurones, some being stimulatory and others inhibitory. Thus, the activity of endocrine neurones at any point in time reflects an interplay between a variety of positive and negative inputs.

What role do neurones arising outside the hypothalamus play in the control of reproduction? Such neurones transmit information relevant to a variety of sensory stimuli (visual, olfactory, etc.; see Fig. 1.1). In contrast, neurones arising within the hypothalamus and forming synapses with endocrine neurones facilitate communication between the various hypo-thalamic nuclei. This permits a finely tuned coordination of the numerous endocrine functions controlled by the brain.

At the level of the synapse, the incoming neurone discharges a specific neurotransmitter substance. This is bound to membrane receptors on the post-synaptic neurone thus modifying its activity, including a change in its rate of firing and discharge of neurotransmitters or hormones from its terminal. Which neurotransmitters control gonadotrophin secretion? Much effort has been spent in identifying the neurotransmitters and attempting to correlate their pattern of release with various reproductive conditions. The release of neurotransmitters, unfortunately, has proven to be extremely difficult to monitor, and we still have much to learn about this important level of control. Nevertheless, a number of experimental approaches point to a role for three monoamines: dopamine and noradrena-line (catecholamines), and serotonin (indoleamine). The hypothalamus is rich in these monoamines. Distinct changes in secretion of LH, FSH and prolactin can be induced by administering them, or by the application of agents that alter their synthesis, metabolism, or the activity of post-synaptic receptors that bind them.

We should recall, at this juncture, that one of these monoamines, dopamine, also functions as a hormone, because it is secreted into the portal system and acts on the anterior pituitary to inhibit the secretion of prolactin. Furthermore, the neurotransmitter functions of dopamine probably include the regulation of GnRH secretion. This points to an overlap in the mechanisms that regulate LH, FSH and prolactin, and may explain why gonadotrophin secretion often tends to be low when prolactin secretion is high.

More recently, a great deal of interest has arisen in another class of compounds, the endogenous opioid peptides. One of these peptide neuro-transmitters, β-endorphin, is found in a high concentration in the hypothalamus and pituitary portal blood. Furthermore, the administration of opioid peptides seems to inhibit the secretion of LH and FSH, and stimulate the secretion of prolactin, whereas opioid antagonists like naloxone can stimulate gonadotrophin secretion.

Finally, the releasing hormones themselves may play limited roles as neurotransmitters. For example, it has been claimed that GnRH may serve as a neurotransmitter in the control of sexual behaviour.

Hormonal regulation

The second level of control over endocrine neurones is provided by blood-borne hormones acting through feedback loops. Although it is known that the brain is a target tissue for hormones, the specific nature of the interaction is not yet clear. For example, gonadal steroids bind to high-affinity receptor proteins in neurones, much as they do in other target tissues where their effects are evoked through the genome and modification of protein synthesis (see Chapters 4, 5 and 6 in Book 7, First Edition). Nevertheless, there is some indication that the more rapid (millisecond) effects of steroids on neurones are mediated by other mechanisms not involving protein synthesis.

Given that both hormones and neurotransmitters modify the activity of endocrine neurones, what is the nature of this interplay? Do hormones modulate the release of neurotransmitters? Do they alter the response of endocrine neurones to neurotransmitters, perhaps by changing excitability thresholds, membrane permeability, or receptors for neurotransmitters? Do hormones function as neurotransmitters, directly generating or inhibiting action potentials? These are crucial questions, but ones for which we have no answers.

Finally, binding sites for the gonadal steroid hormones are not confined to the hypothalamus but are widely distributed throughout the brain. Combined with evidence that implantation of steroids into discrete neural sites outside the hypothalamus can have profound effects on reproduction, the wide distribution of binding sites suggests that the actions of gonadal hormones are not merely restricted to the hypothalamus. Clearly, these steroid hormones also act on extrahypothalamic neurones which, in turn, link up with endocrine neurones in the hypothalamus. Regardless of where the gonadal hormones act in the brain, the end result is a change in the rate of discharge of the hypothalamic hormones into the portal circulation and a modulation of pituitary hormone secretion.

Control of gonadotrophin secretion

Tonic and surge modes of secretion

It is generally accepted that the secretion of LH and FSH is controlled by two functionally separable but superimposable regulatory systems – the tonic system and the surge or cyclic system (Fig. 1.8). The tonic system produces the ever-present basal levels of circulating pituitary hormones, which promote the development of both germinal and endocrine elements of the gonads. The surge system, in contrast, operates rather acutely, generally being evident for only 12–24 hours in each reproductive cycle of the female. The surge mode of secretion is responsible for a massive (often several hundred-fold) increase in circulating gonadotrophin, the primary function of which is to cause ovulation in the female (see Chapters 5 and 6).

It is the surge mode of gonadotrophin secretion that endows the female with the capacity to cycle. In many species (such as the rat, mouse, guinea pig and sheep), the surge system is capable of functioning only in the female. Although this system is present in the developing fetus of *both* sexes, it is rendered permanently inoperative in males by the action of testicular hormones secreted during a critical stage of intra-uterine or early postnatal development (see Chapter 3 of Book 2). In such species, this process of sexual differentiation occurs in the brain itself, not in the anterior pituitary, which remains sexually undifferentiated.

The existence of a tonic system in both sexes and a surge system only in females, is in harmony with the concept that the tonic and surge regulatory centres reside in different regions of the hypothalamus. The clearest demonstration of such a spatial separation comes from work on the rat (Fig. 1.9). In this species, the tonic mode of gonadotrophin secretion is governed by endocrine neurones whose cell bodies are located in the medial basal hypothalamus, specifically in the arcuate nucleus. The gonadotrophin surge, however, requires inputs from more anterior regions, the preoptic area and suprachiasmatic nuclei. It is the anterior regions that

Fig. 1.8. Schematic representation of two different modes of secretion of FSH and LH. Note that tonic secretion stimulates gonadal development whereas the surge causes ovulation.

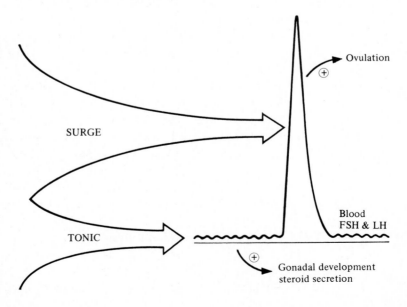

appear to undergo sexual differentiation; in fact, recent evidence points to a large structural difference in the preoptic area of male and female rats (see Book 2, Chapter 3).

Another interesting feature of the LH surge system in rodents is that its operation is coupled to the light–dark cycle. Thus, the LH surge normally occurs at a fixed time of the day (late afternoon). This timing of the LH surge appears to originate in the general region of the suprachiasmatic nucleus, a hypothalamic area believed to contain an important component of the biological clock.

To what extent do these characteristics of the LH surge system apply to other species? It is likely that considerable differences exist. In the rhesus monkey, for example, the LH surge system is not coupled to the light–dark cycle, and it is not sexually differentiated. LH surges are not normally seen in male monkeys, but this is due to the fact that the hormones secreted from the testes of the adult are not compatible with the induction of the LH discharge. The surge system itself is fully competent in the male. In another primate, the marmoset monkey, the LH surge system is present

Fig. 1.9. Diagram illustrating the anatomical location of the centres controlling the tonic and surge modes of gonadotrophin (Gn) secretion in rodents and monkeys. Note that the surge system is obliterated in male rodents (which have a *larger* preoptic area than females) but not in monkeys. POA, preoptic area; SCN, supra-chiasmatic nucleus; ARC, arcuate nucleus; ME, median eminence; OC, optic chiasma; AP, anterior pituitary. GnRH, gonadotrophin releasing hormone.

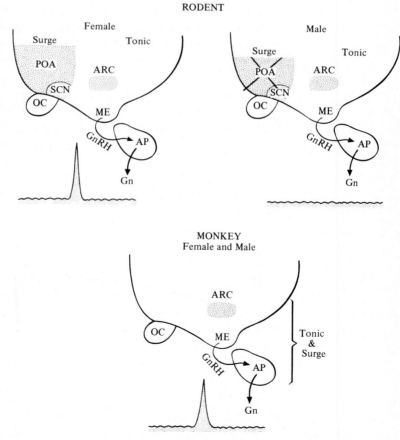

in both sexes, and even in man the LH surge may be induced in males if hormonal conditions are right.

Species differences in the LH surge system may reflect the relative importance of the anterior hypothalamic regions in generating the surge. For example, the preoptic area and suprachiasmatic nuclei, which are clearly needed for this event in rodents, can be obliterated in the rhesus monkey without disrupting the surge mode of gonadotrophin secretion. In fact, it has not even been possible to separate the tonic and surge systems anatomically in the monkey. Thus, the medial basal hypothalamus (arcuate nucleus), median eminence, and anterior pituitary gland seem able to act as a self-contained unit, with all the essential neuroendocrine elements for both modes of secretion (Fig. 1.9).

Finally, in considering species differences in the gonadotrophin surge, it is important to remember that in some animals (like the rabbit, cat, ferret, and camel) this event is normally triggered by the act of copulation. In these species, which are referred to as induced or reflex ovulators, the stimulation of nerve endings in the vagina and cervix activates a neuro-endocrine reflex which includes the transmission of impulses up the spinal cord to the hypothalamus and the activation of GnRH-producing neurones. In most species, however, the gonadotrophin surge occurs independently of copulation. In such animals, which are called spontaneous ovulators, the surge system is activated solely by the stimulatory influence of gonadal hormones. This brings us to a consideration of the feedback controls.

Feedback control of gonadotrophin secretion
Most hormones in the body are part of homeostatic feedback loops by which each hormone regulates its own rate of secretion within well-defined limits. Three operational levels of hormonal feedback have been proposed for the hypothalamo-pituitary–gonadal system: long-loop feedback, short-loop feedback, and ultra-short-loop feedback (Fig. 1.10). The best documented of these is the long-loop feedback, in which the hormones secreted from the gonads exert their regulatory effects at the level of the anterior pituitary gland and/or the central nervous system (level 1 in Fig. 1.10). Short-loop feedback (level 2) refers to an action of pituitary hormones to regulate their own secretion, perhaps by acting within the hypothalamus to which they are delivered by back-flow of blood up the pituitary stalk. Ultra-short-loop feedback refers to a neural action of hypothalamic hormones to regulate their own secretion (level 3).

Most endocrine feedback loops are negative (inhibitory) because the hormone in question effects a decrease in its own rate of secretion. Some hormonal feedbacks, however, are stimulatory (positive) because, up to a certain point, a hormone may stimulate its own release. The tonic mode of gonadotrophin secretion is governed by a negative feedback action of gonadal steroids; removal of the gonads, therefore, produces a large increase in tonic LH and FSH secretion (example in Fig. 1.11). The surge system, in contrast, is regulated by a positive feedback action of gonadal steroids, at least in spontaneous ovulators.

Which particular gonadal steroids play feedback roles and where do they act in the hypothalamo-pituitary axis? Three types of gonadal steroids (oestrogen, progesterone and androgen) play major negative feedback

Fig. 1.10. Illustration of three levels of hormonal feedback proposed for the hypothalamo-pituitary–gonadal axis: (1) long-loop feedback; (2) short-loop feedback; (3) ultra-short-loop feedback. GnRH, gonadotrophin releasing hormone; AP, anterior pituitary; PP, posterior pituitary.

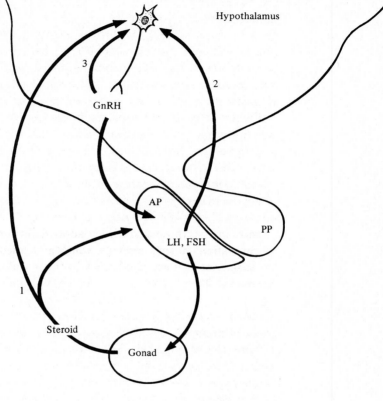

Fig. 1.11. *Left*: Example of long-loop negative feedback between oestrogen (OE) and LH. Moving clockwise from the top of the loop, an increase in OE causes a decrease in tonic LH secretion. Since LH stimulates OE secretion, the drop in LH causes a decline in OE. This removes the inhibition on LH which therefore increases. The rising titre of LH then stimulates ovarian oestrogen secretion, and so on. *Right*: Representation of the rise in serum LH that occurs when the negative feedback loop is broken by ovariectomy (OVX).

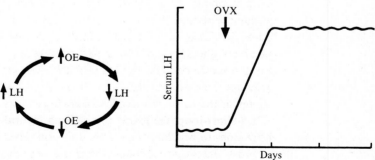

roles. Oestrogen also serves a positive feedback function because it elicits the gonadotrophin surge in females. These steroids may act within the brain to modify the rate of secretion of the releasing–inhibiting hormones; they may act on the anterior pituitary gland to alter its response to the hypothalamic hormones, or they may act at both sites. Which site prevails varies with the species, the steroid, and whether the feedback is positive or negative.

The hypothalamic pulse generator

Perhaps one of the more fascinating aspects of the neuroendocrine regulation of reproduction is the pulsatile nature of hormone secretion. This property of the system first became evident around 1970, initially from classical studies in the rhesus monkey by Ernst Knobil and his colleagues, and extended shortly thereafter to a wide variety of species. In general, the tonic mode of gonadotrophin secretion (most notably LH) does not proceed at a steady rate. Rather, it is characterized by discrete bursts of secretion separated by periods when there is relatively little or no secretion. This pulsatile pattern is most evident in gonadectomized animals in which the tonic control system is free of negative feedback suppression (Fig. 1.12).

Discovery of the episodic nature of gonadotrophin release has led to the concept that the tonic mode of secretion is controlled by an oscillator, or pulse generator, which resides in the central nervous system. It has recently been confirmed, by Gary Jackson's group in Illinois and Iain Clarke's group in Melbourne, that this oscillator generates rhythmic quantal discharges of GnRH from endocrine neurones which, in turn, drive pulses of gonadotrophin secretion (Fig. 1.12, lower part).

Where is the pulse generator located and how does it work? The pulse generator appears to be contained within the medial basal region of the hypothalamus in the few species in which it has been localized. Not much is known about how it works, largely because of methodological difficulties in pin-pointing and studying the small parvicellular neurones that secrete GnRH. The magnocellular endocrine neurones which release the posterior pituitary hormones, however, also operate in pulsatile fashion, and a great deal more is known about them (see Chapter 2). Based on findings in magnocellular neurones, we could infer that the pulsatile pattern of secretion arises from a rhythmic and synchronous firing of entire populations of endocrine neurones. This produces a synchronous quantal discharge of hormones stored in the hypothalamic nerve terminals.

It has recently become recognized from studies in the rhesus monkey, that the pulsatile nature of GnRH secretion is obligatory for normal gonadotrophin secretion, and hence for reproductive success. In female monkeys whose endogenous GnRH secretion has been abolished by lesions of the arcuate nucleus, an episodic pattern of replacement of GnRH reinstates normal gonadotrophin secretion and menstrual cycles, whereas constant infusions of GnRH are ineffectual.

Control of the pulse generator. The activity of the pulse generator is normally modified by a wide variety of inputs which convey information about both the internal and external environment. For example, the post-castration rise in tonic gonadotrophin secretion is caused by a large increase in both frequency and amplitude of LH pulses. The negative feedback actions of gonadal hormones, therefore, must be effected through the pulse-generating system.

In the presence of the gonads, the frequency and amplitude of gonadotrophin pulses normally undergo big fluctuations. These changes are

Fig. 1.12. Illustration of the pulsatile hormone secretion in an ovariectomized female, and the postulated neural basis for pulsatile secretion of pituitary hormones. AP, anterior pituitary, GnRH, gonadotrophin-releasing hormone; LH, luteinizing hormone. (Adapted from R. L. Goodman and F. J. Karsch. In *Biological Clocks in Seasonal Reproductive Cycles*, Colston Papers No. 32, ed. B. K. Follett and D. E. Follett. John Wright; Bristol (1981).)

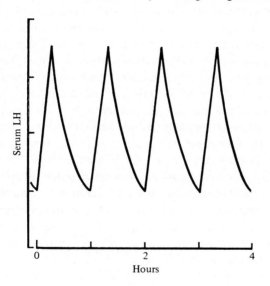

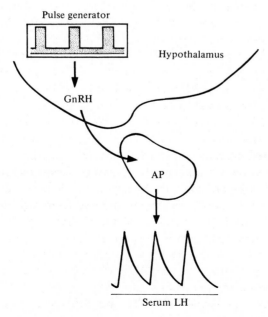

attributable, in part, to variations in particular gonadal hormones acting at different sites in the hypothalamo-pituitary axis (Fig. 1.13). In some species, for example, the frequency of LH pulses seems to be inhibited by progesterone in the female and and testosterone in the male, implying a neural site for the feedback effects of these steroids to slow the pulse generator. LH pulse amplitude, on the other hand, is reduced in both sexes by oestradiol, acting on the pituitary to decrease its response to the oscillations of GnRH delivered in the portal blood. (Amplitude of LH pulses may also be decreased by a reduction in the amount of GnRH discharged in each bolus; it is not yet known whether oestrogens also act in this way.)

Profound changes in the pulse generator also occur at puberty, during the course of the year in seasonal breeders (see Chapter 3), and in response to a number of acute external stimuli originating from visual, olfactory and tactile cues. One of the most striking examples of the latter is the LH pulse pattern in male mice during exposure to a female (Fig. 1.14). Art Coquelin and Frank Bronson have observed that initial exposure to a female mouse promptly elicits an LH pulse in the male. Repeated exposure to the same female evokes additional LH pulses, but of progressively decreasing amplitude until the response barely occurs. At this point, exposure to a *new* female restores the response, producing a full-blown LH pulse in the

Fig. 1.13. Control of pulsatile LH secretion by gonadal hormones. Note differential modulation of amplitude and frequency. AP, anterior pituitary; PP, posterior pituitary; GnRH, releasing hormone; LH, luteinizing hormone.

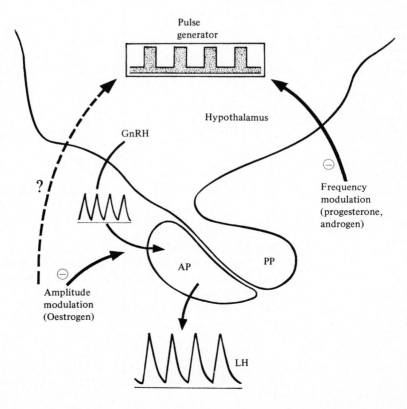

male. This fascinating observation, which is linked to olfactory cues arising from the female urine, demonstrates not only that social–olfactory stimuli can impinge upon the pulse generating system for LH, but also that the system is susceptible to habituation. Studies by David Lindsay and his colleagues in Western Australia have also shown that exposure of the ewe to the smell of a ram will increase her frequency of pulsatile LH discharge, hence providing a clue how introduction of the ram can hasten and synchronize the onset of oestrus in a flock of anoestrous ewes. These exciting observations open up a whole new dimension to our understanding of the neuroendocrine regulation of gonadotrophin secretion.

Since the first discovery of the vascular link between the hypothalamus and anterior pituitary gland in the 1930s and 1940s, we have come a long way in understanding the means by which the central nervous system regulates the activity of the gonads. Many intricate details of the complex network of information transfer outlined in Fig. 1.1 have been clarified, but many pieces of the puzzle still need to be put in place. It used to be said that the anterior pituitary is the conductor of the endocrine orchestra; if that is so, then it is the hypothalamus that writes the score, in response to feedback from the audience.

Fig. 1.14. Control of pulsatile LH secretion in male mice by exposure to the female. Note that exposure to a female (♀1) promptly elicits an LH pulse, but the size of the pulse decreases with repeated exposure to the same female. Exposure to a new female (♀2) can restore the initial response. (Adapted from A. Coquelin and F. H. Bronson. *Science*, **206**, 1099 (1979).)

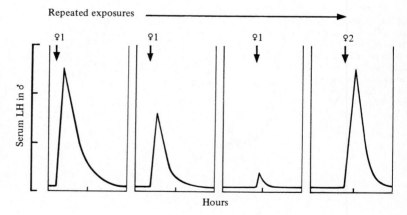

Suggested further reading

Endocrine mechanisms governing transition into adulthood in female sheep. D. L. Foster and K. D. Ryan. *Journal of Reproduction and Fertility*, *Supplement* **30**, 75–90 (1981).

Seasonal reproduction: a saga of reversible fertility. F. J. Karsch. *The Physiologist*, **23**, 29–38 (1980).

The neuroendocrine control of the menstrual cycle. E. Knobil. *Recent Progress in Hormone Research*, **36**, 53–88 (1980).

Seasonal breeding: nature's contraceptive. G. A. Lincoln and R. V. Short. *Recent Progress in Hormone Research*, **36**, 1–43 (1980).

Can the pituitary secrete directly to the brain? (Affirmative anatomical evidence.) R. M. Bergland and R. B. Page. *Endocrinology*, **102**, 1325–38 (1978).

Neuroendocrinology of reproduction. A Symposium. *Biology of Reproduction*, **20**, 1–27 (1979).

Neuroendocrine control of reproduction. Symposium Report No. 15. *Journal of Reproduction and Fertility*, **58**, 493–554 (1980).

Handbook of Physiology. Section 7: *Endocrinology*, volume 4, Parts 1 and 2, *The Pituitary Gland and Its Neuroendocrine Control*. Ed. E. Knobil and W. H. Sawyer. American Physiological Society; distributed by Williams and Wilkins and Co.; Baltimore (1974).

Neuroendocrinology. Ed. D. T. Krieger and J. C. Hughes. Sinauer Associates, Inc.; Sunderland, Mass. (1980).

The hypothalamic pulse generator: a key determinant of reproductive cycles in sheep. R. L. Goodman and F. J. Karsch. In: *Biological Clocks and Seasonal Reproductive Cycles*, Colston Papers No. 32, p. 223. Ed. B. K. Follett and D. E. Follett. John Wright; Bristol (1981).

The Nobel Duel. N. Wade. Doubleday; New York (1981).

2

The posterior pituitary

DENNIS W. LINCOLN

The posterior lobe of the pituitary gland is part of the central nervous system, and is primarily composed of the axon terminals of neurones whose cell bodies are situated some distance away in two well-defined hypothalamic nuclei (Fig. 2.1). These neurones exhibit the electrophysiological properties of nervous tissue, though at the same time they produce and then release as a function of their electrical activity large quantities of either oxytocin or vasopressin. For this reason, Chapter 2 must be concerned not only with the posterior pituitary but also with the hypothalamic neurones that contribute to its structure and function.

Oxytocin gained early recognition for its ability to promote the delivery of the young, and was named accordingly (Gr. *oxys*, swift, + *tokos*, childbirth). Certainly this is the case with some species. The domestic rabbit is a good example; an injection of 50 mU oxytocin within 48 hours of term causes the birth of the entire litter within 10 minutes and in a manner that is indistinguishable from that which occurs naturally. Furthermore, the

Fig. 2.1. A section parallel to the mid-line in a mammalian hypothalamus and pituitary gland, diagrammatically illustrating the course taken by the axons of magnocellular neurones from the paraventricular and supraoptic nuclei to the posterior pituitary.

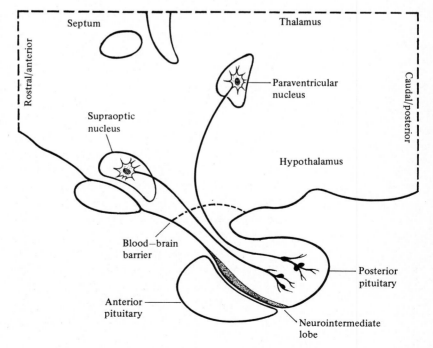

doe seems able to control the time of birth through the release of her own oxytocin. The picture is less clear for many other species, including ourselves, though oxytocin may play a more important role than is generally recognized. Here the key may be the rapid development of oxytocin receptors within the uterus on which low levels of oxytocin already in the circulation may act, rather than the release of increased amounts of hormones. Oxytocin assumes a second role in the days and weeks following birth, and one that is quite indispensable for the natural survival of most young. At this time it stimulates the contraction of the myoepithelial cells investing the alveoli of the mammary gland, and milk stored in the alveoli, more than 80 per cent of the total amount in most species, is ejected into the larger mammary ducts from where it can be withdrawn by the sucking of the young. For several years there has been tentative evidence to suggest that oxytocin might play a role in ovulation, in the tubal transport of ova, and in the movement of spermatozoa through the male and female reproductive tracts, but further study is required. One exciting possibility to emerge recently is that this hormone, acting on oestrogen-induced oxytocin receptors, may cause the release of prostaglandin $F_{2\alpha}$ within the uterus and thereby evoke the regression of the corpus luteum during the oestrous cycle (see Chapter 5). By contrast vasopressin has no direct reproductive functions. Vasopressin, or antidiuretic hormone as it is sometimes more appropriately known, acts on the kidney to promote the resorption within the collecting ducts of more than 99 per cent of the fluids filtered from the blood, and so serves to maintain the fluid balance of the body.

The importance of oxytocin in labour and lactation, and its possible role in several other key reproductive events, is one reason for including a chapter on the posterior pituitary in the Second Edition of 'Reproduction in Mammals'; however there is another reason which may not be so immediately obvious to reproductive biologists. The posterior pituitary and its innervation from the hypothalamus provide a model, perhaps the best model, through which to explore many of the fundamental principles underlying the organization of the neuroendocrine system. One does not have to look far for the reason. The discrete localization of the peptide-producing cells and that of their secretory terminals at separate but relatively accessible sites permits the independent investigation of the processes of peptide synthesis, transport and secretion. Likewise, this anatomical arrangement allows the electrophysiologist to separate the generation of the action potential from its conduction and from the events associated with the secretion at the nerve terminal. Studies have capitalized on this elegant arrangement for the last 15 years, but recently new impetus has been added by the realization that the processes of peptide secretion displayed by the neurones innervating the posterior pituitary may apply to the release of the other regulatory peptides throughout the nervous system. These include substance P, TRH, GnRH and the enkephalins. It would now seem that so-called endocrine cells, even those in the anterior pituitary gland, may generate electrical potentials that determine secretion.

Organization of the magnocellular neurosecretory system

Most of the neurones that produce oxytocin and vasopressin are found in the paired supraoptic and paraventricular nuclei of the hypothalamus (Fig. 2.1), though a significant number of cells are also seen in scattered groups between these nuclei. In this context, the term *nucleus* merely relates to an aggregation of neurones – it does not imply any particular function. These peptide-producing neurones are often referred to as the magnocellular neurones because their cell bodies, at 30 μm diameter, are 2–3 times larger than most other hypothalamic neurones. This is of course a relative judgement: the Betz cells of the cerebral cortex have diameters greater than 100 μm.

Current evidence based largely on immunohistochemistry and backed up by biochemical and electrophysiological studies strongly suggests that oxytocin and vasopressin are produced in separate neurones. There are reports to indicate some regional localization of the oxytocin- and vasopressin-producing neurones within each nucleus, though there is general agreement that both cell types are extensively intermingled. Such an arrangement is very much in accord with the view that the supraoptic and paraventricular nuclei have evolved from the single preoptic nucleus of premammalian vertebrates. However it would be wrong to imply that the magnocellular nuclei are identical. The paraventricular nucleus of the rat, for example, appears to be more involved in the release of oxytocin than of vasopressin, though there are still more oxytocin-producing cells in the supraoptic nuclei due to the fact that these nuclei are three or four times larger. In addition the magnocellular portion of the paraventricular nucleus is surrounded by a number of parvicellular nuclei and some neurones within these nuclei also produce oxytocin and vasopressin. These parvicellular neurones have axons running to such diverse places as the median eminence, the spinal cord and the amygdala.

Little is known of the organization within the magnocellular nuclei, and it is easy to generate many more questions than answers. Each neurone displays upwards of 5000 synapses on its cell body or dendrites, and through these may receive information from 1000 other neurones. These synapses are typical of those found elsewhere in the brain, and depend on noradrenaline and acetylcholine, for example, as neurotransmitters. There is no single prominent tract of fibres impinging on the magnocellular nuclei, but rather the sensory input appears to come from many parts of the nervous system. One fact though is perhaps worthy of note in this context: up to 60 per cent of the synapses within the magnocellular nuclei fail to degenerate after the nuclei have been surgically isolated from the rest of the brain. This suggests that many of the synapses may originate from neurones within the nuclei themselves or in adjacent tissue. Electrophysiological evidence to be presented later indicates that oxytocin cells are linked into an electrical syncytium, that vasopressin cells have an inherent mechanism for phasic discharge, and that both cell types may be involved in some form of electrophysiological feedback, possibly via interneurones. There is, however, scant anatomical evidence to account for any of these well-documented phenomena.

The axons of the magnocellular neurones, less than 1 μm in diameter and unmyelinated, pass down through the basal hypothalamus and pituitary stalk to enter the posterior pituitary, a distance of some 3–4 mm in the rat. In the process these axons cross the 'blood–brain barrier', and that could be rather more important than has hitherto been appreciated. The endothelial cells lining the capillaries of the brain, unlike those elsewhere in the body, are joined by tight junctions; this prevents large molecular weight peptides in the peripheral circulation (such as luteinizing hormone and β-endorphin) from entering the brain other than in minute amounts, whereas small lipid-soluble molecules (like gonadal steroids) may do so quite freely. The posterior pituitary, the median eminence and certain other structures surrounding the third ventricle of the brain are outside the blood–brain barrier, and are therefore exposed to the actions of circulating peptides.

Once within the posterior pituitary, the axons of the magnocellular neurones exhibit many dilatations and divide profusely before terminating on capillaries. These dilatations or Herring bodies (named in honour of Theodore Herring who first described them in 1908 whilst working as a Research Fellow in Edinburgh) contain stores of oxytocin and vasopressin. The posterior pituitary also has structural elements of its own, known as

Fig. 2.2. A family tree depicting the evolution and structure of the posterior pituitary peptides of mammals. Gene duplication permits mammals to produce simultaneously two (or more) posterior pituitary peptides, whilst point mutations result in the substitution of one amino acid for another. Possible changes in the nucleotide sequences that could account for these mutations are shown, and the substituted amino acids have been ringed for identification. AGA, for example, represents a nucleotide sequence of adenosine–guanosine–adenosine, and codes for the amino acid arginine.

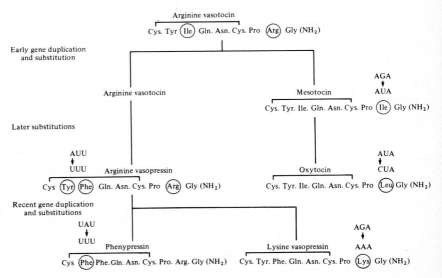

pituicytes. These cells appear to be very active in fetal life and it has been suggested that they produce chemotaxic factors that serve to attract the axons of the magnocellular neurones as they grow down from the hypothalamus. No function has been firmly established for the pituicyte in adult life, though lysosomal activity in the pituicytes is greatly increased under conditions of enhanced secretion from the magnocellular nerve terminals.

Structure of the posterior pituitary peptides

Oxytocin and vasopressin were first synthesized in 1951, and for this first synthesis of a peptide hormone (or neuropeptide) Vincent du Vigneaud of Cornell University Medical College was awarded the Nobel Prize for Chemistry in 1955. Both hormones contain nine amino acids with the cysteine residues in positions 1 and 6 joined by a disulphide bridge to produce a ring configuration (Fig. 2.2). Seven of the amino acids are common to both peptides, and this results in a considerable interaction of the two peptides at their respective receptors. Vasopressin, for example, will promote milk ejection from the mammary gland, though its potency in the rat is only 25 per cent that of oxytocin.

Several hundred analogues of the posterior pituitary hormones have been produced in an attempt to uncover the principles underlying the biological actions of the natural peptides, as well as to search for peptides with useful therapeutic or pharmacological properties. Few of the derived peptides have greater milk-ejecting activity than oxytocin, whilst single amino acid substitutions frequently result in substantial loss of biological activity or generate antagonists that bind to the receptors but fail to

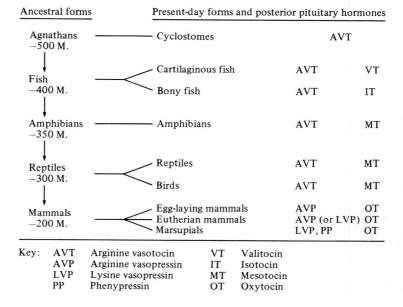

Fig. 2.3. Occurrence of posterior pituitary peptides in modern-day vertebrates. The time when the ancestral forms of these various vertebrates first appeared is given on the left in millions of years (M).

Ancestral forms	Present-day forms and posterior pituitary hormones		
Agnathans −500 M.	Cyclostomes	AVT	
Fish −400 M.	Cartilaginous fish	AVT	VT
	Bony fish	AVT	IT
Amphibians −350 M.	Amphibians	AVT	MT
Reptiles −300 M.	Reptiles	AVT	MT
	Birds	AVT	MT
Mammals −200 M.	Egg-laying mammals	AVP	OT
	Eutherian mammals	AVP (or LVP)	OT
	Marsupials	LVP, PP	OT

Key:	AVT	Arginine vasotocin	VT	Valitocin
	AVP	Arginine vasopressin	IT	Isotocin
	LVP	Lysine vasopressin	MT	Mesotocin
	PP	Phenypressin	OT	Oxytocin

activate them. The ring configuration and the *N*-terminal group of oxytocin are both essential for biological activity, while modifications at positions 2, 3, 4 and 5 within the ring produce a range of less active products. Changes at position 8 are of particular interest because many of these peptides, occurring in fish, birds and reptiles, are quite active in mammals.

The evolution of the posterior pituitary hormones has been the subject of considerable debate owing in no small part to the fact that there is no way of measuring what posterior pituitary peptides were produced by premammalian forms. One is faced with having to make deductions based on an analysis of a minute percentage of present-day vertebrates. Most of the species that have been studied, with the exception of the cyclostomes (lampreys), have two posterior pituitary peptides resembling oxytocin and vasopressin, and occasional species, including some marsupials, have three (Fig. 2.3). Secondly, arginine vasotocin occurs in a great diversity of species, and from this one is tempted to conclude that arginine vasotocin was the ancestral peptide form which the oxytocin and vasopressin of Eutherian mammals were derived. Most mammals produce arginine vasopressin, though a few such as the pig and its relatives produce lysine vasopressin. It would require few genetic changes to convert arginine vasotocin to the peptides found in mammals. Early in evolution, and perhaps more recently in some species, there must have been gene duplication to permit two or more posterior pituitary peptides to co-exist in the same species. Thereafter oxytocin and vasopressin (in either its arginine or lysine forms) can be derived by point mutations in single DNA codons. Fig. 2.2 gives one possible sequence by which the mammalian peptides may have been derived, and one set of codon changes to account for the observed differences in amino acid sequence.

Interestingly, oxytocin is found in all mammals, including the egg-laying echidna and platypus, but not in birds, reptiles or amphibians. This suggests that the switch from mesotocin to oxytocin (see Fig. 2.2) must have occurred in parallel to the evolution of the mammary gland, but there is no reason to interpret these two developments as related events. Mesotocin has 75 per cent of the milk-ejecting activity of oxytocin and would have served as an excellent alternative.

Biosynthesis of oxytocin and its prohormone

The magnocellular neurones are specialized for the production and storage of large amounts of oxytocin and vasopressin, and in common with other endocrine cells the active peptides are produced by the cleavage of a large molecular weight precursor or prohormone. A glycosylated peptide of 166 amino acids and a molecular weight of about 20 000 is synthesized within the rough endoplasmic reticulum of the cell body (Fig. 2.4). This is then moved to the Golgi apparatus and is there packaged into membrane-bound granules of about 100–150 nm diameter (Fig. 2.5). Glycosylation may be important in the process of granule formation, because granule formation ceases when *N*-glycosylation is prevented by treatment of rats with the enzyme inhibitor tunicamycin. Some peptide probably remains in an extragranular form, but how much is a frequent matter of debate, and the situation is complicated by the possibility that many technical procedures may themselves rupture the granule membranes. Presumably the packaging of the peptide into granules facilitates its transport, storage and quantal release. At the same time, it could relieve the cell from what might otherwise be an intolerable level of one peptide, because in some respects the contents of these neurosecretory granules are 'extracellular' even before they are released.

Fig. 2.4. A stylized view of a hypothalamic magnocellular neurone involved in hormone biosynthesis. The stages in hormone production, from the uptake of precursors by the hypothalamic neurone to the storage of the final products in the nerve terminals of the posterior pituitary, are given on the right. Axonal transport of peptides is a bidirectional process, and on the left is illustrated the possibility that granule membrane, recaptured after the exocytosis of the granule contents, may be returned by retrograde transport to the cell body.

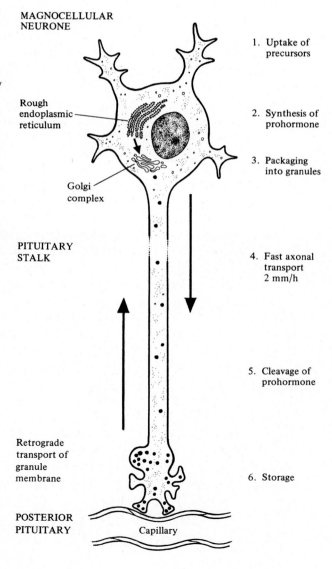

MAGNOCELLULAR
NEURONE

1. Uptake of precursors

Rough endoplasmic reticulum

2. Synthesis of prohormone

3. Packaging into granules

Golgi complex

PITUITARY STALK

4. Fast axonal transport 2 mm/h

5. Cleavage of prohormone

Retrograde transport of granule membrane

6. Storage

POSTERIOR PITUITARY

Capillary

One component of the prohormone, called the neurophysin, was identified in pituitary extracts several decades ago, and was for a long time thought to function as a 'carrier protein' for the transport of the posterior pituitary hormones. Today the neurophysin appears to be physiologically redundant, though it is a useful peptide through which to investigate the posterior pituitary, being produced and released in parallel to its respective hormone. The rat has two major neurophysins (A and B, using the nomenclature proposed by Brian Pickering in Bristol), and one present in much smaller amounts (C) (Fig. 2.6*b*). The chance discovery of a mutant strain of rats in Brattleboro, Vermont, by Heinz Valtin, has shed much light on the association between the hormones and the neurophysins. The homozygous Brattleboro rat produces no vasopressin and displays diabetes

Fig. 2.5. (*a*). A transmission electronmicrograph of a section from the posterior pituitary of a rat to illustrate the accumulation of membrane-bound neurosecretory granules within a Herring body (HB). Scale 1 μm. (*b*). A freeze–fracture image of an axon of a magnocellular neurone to illustrate changes indicative of exocytosis of neurosecretory granules. The fracture has exposed a large area of the inner leaflet (P-face) of the plasma membrane (adjacent to the cytoplasm) and part of the axon's cytoplasm which is filled with neurosecretory granules (G). Several exocytotic openings are apparent on the P-face, delineated by dotted lines, and one such opening contains granule core material (arrow). Scale 0.2 μm. (*c*). A transmission electronmicrograph of a neurohypophysial terminal from a rat that has been water deprived to increase the release of posterior pituitary hormones. A round mass (arrow) of electron density similar to the core of a neurosecretory granule fills an invagination in the cell membrane. Scale 0.2 μm. (*d*). A freeze–fracture image of an exocytotic invagination corresponding to the section illustrated in (*c*). In this replica the fracture passed along the P-face of the plasma membrane into the cytoplasm of the same axon, thus revealing the fused granule and plasma membranes. A cross fracture of the granule core fills the exocytotic opening. Scale 0.2 μm. (Micrograph (*a*) was kindly provided by Peter Heap of the Department of Anatomy, University of Bristol, and micrographs (*b*)–(*d*) by Dennise Theodosis of the Institut National de la Santé et de la Récherche Medicale, Bordeaux. For further details see: D. T. Theodosis, J. J. Dreifuss & L. Orci. *J. Cell Biol.* **78**, 542–53 (1978).)

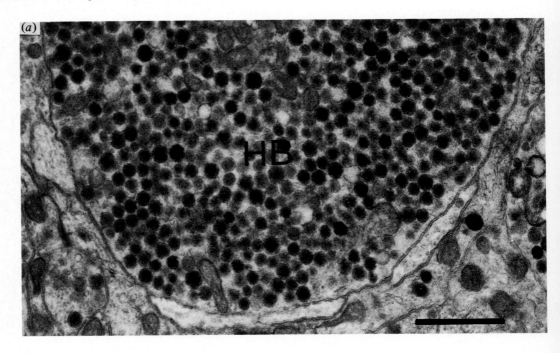

(*a*)

insipidus, excreting 70 per cent of its body weight/day in urine compared with 3 per cent for a normal rat. In addition, the Brattleboro rat is deficient in neurophysin A, and so presumably this forms part of the prohormone from which vasopressin is cleaved. Neurophysin B, with a somewhat different amino acid sequence, is probably associated with the production of oxytocin, whilst neurophysin C may be a breakdown product of B. An alternative numerical nomenclature has been proposed where neurophysin I is the peptide with the fastest electrophoretic mobility. This has resulted in some confusion; for example, bovine neurophysins I and II appear to be associated with oxytocin and vasopressin, respectively, whilst it is the other way round for porcine neurophysins I and II.

Our understanding of the relationship between the active hormone and

its precursor has recently been enormously advanced by the sequencing of the entire precursor molecule by Dietmar Richter and his colleagues. Their approach to the problem was indirect, and involved the insertion of the genes for the production of the precursor into a bacterial plasmid. This was then cloned and the large amount of mRNA so produced was used to determine the amino acid sequence. The precursor consists of four components connected by linkage groups as follows: signal peptide, hormone, neurophysin and glycopeptide (Fig. 2.7). The signal peptide is probably involved in biosynthetic processes within the endoplasmic reticulum, and is cleaved from the main structure before packaging occurs in the Golgi apparatus. The size of the carbohydrate fraction attached at position 115 is in the range of 2000–4000 daltons.

Fig. 2.6. Studies of peptide biosynthesis in the magnocellular neurones of the rat. In part (*a*) the posterior pituitary peptides have been labelled by the intracisternal (or intraventricular) injection of [³⁵S]cysteine. Both oxytocin and vasopressin contain two cysteine residues (see Fig. 2.2), and the injection of the precursor into the ventricular system of the brain maximizes the amount reaching the hypothalamus. The first arrival of labelled peptide in the posterior pituitary gives evidence of the speed of peptide synthesis and transport, whilst the loss of activity in the weeks following provides an index of turnover. (Data from G. Burford, C. W. Jones and B. T. Pickering. *Biochem. J.* **124**, 809–13 (1971).) Part (*b*) depicts the separation of the neurophysin components of the prohormone by polyacrylamide-gel electrophoresis. The key to this separation was the addition of 1 μg/ml of bromophenol blue to the buffers; this bound to the neurophysins, changed their charge and caused them to separate during electrophoresis. Three peaks were observed with extracts from the posterior pituitary taken 24 h after the intracisternal administration of [³⁵S]cysteine; these represented neurophysins A, B and C. Peak A was missing in the homozygous Brattleboro rat, which produces no vasopressin. So peak A probably represents the neurophysin component of the vasopressin precursor, whilst B and C are related to the production of oxytocin. (Data from G. Burford and B. T. Pickering. *Biochem. J.* **136**, 1047–52 (1973).)

(*a*)   Transport of labelled precursor

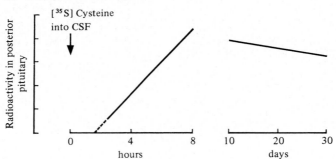

(*b*)   Polyacrylamide-gel separation of neurophysins

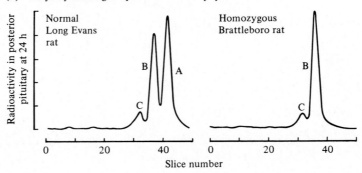

The neurosecretory granules are rapidly transported to the posterior pituitary along the axons of the magnocellular neurones. Labelled peptide appears in the posterior pituitary within 3 h of the injection of a labelled precursor (^{35}S-cysteine) into the ventricles of the rat brain (Fig. 2.6). This indicates a conduction rate of 1–3 mm/h, or a rate 50 times faster than the normal rate of flow of axoplasm down a nerve fibre. One suggestion is that transport involves a form of saltatory (leaping) movement of the granule along the outside of the neurotubules. Certainly, transport is arrested when these tubules have been disrupted by treatment with colchicine, whereas axoplasmic flow is not.

During transport from the hypothalamus to the posterior pituitary the precursor is further cleaved by enzymes, possibly present within the neurosecretory granules, to produce the active hormones. The basic amino acids at positions 11 and 12, lysine and arginine, are important because it is at such sites that tryptic and carboxypeptidase-like enzymes are likely to act and, in common with what is known of the cleavage of other precursor peptides (e.g. pro-opiocortin), the glycine so exposed may function as an amide donor for the terminal amidation of the active hormone (see (Fig. 2.2). Cleavage sites are also seen between the neurophysin and glycopeptide fractions, and a number exist within the glycopeptide fraction itself. This could well result in the production of more fractions

Fig. 2.7. Schematic representation of the precursor for vasopressin based on the nucleotide sequence of mRNA produced by a clone of bacterial plasmids into which cDNA for the precursor, extracted from bovine hypothalamus, had been inserted. The amino acid sequence is numbered from the first cysteine residue of vasopressin; a carbohydrate moiety of 2000–4000 molecular weight (C) is attached at position 115. The section of the precursor containing the active peptide is shown on an expanded scale in the lower part of the figure, and the amino acid sequence illustrated. Cleavage of the signal peptide occurs before granule formation in the Golgi apparatus; nothing is known of the enzymes involved. During axonal transport the remaining peptide is further cleaved in the region of the basic amino acid residues, Lys and Arg at positions 11 and 12, by trypsin- and carboxypeptidase-like enzymes. The glycine residue that is exposed by this cleavage serves as an amide donor for the amidation of the terminal amino acid of the active hormone. (Full details of the precursor sequence and of the methods involved in its analysis are given in H. Land, G. Schütz, H. Schmale and D. Richter. *Nature*, **295**, 299–303 (1982).)

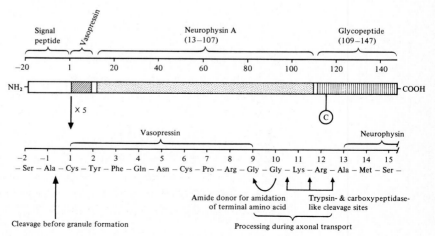

than have been recognized hitherto, and some of these could be biologically active. Oxytocin has been detected in other parts of the nervous system by immunohistochemical procedures, and it is quite plausible to suggest that at these sites other parts of the precursor might be active. Alternatively, by regulating cleavage different cell types could process the common precursor into different active fragments.

Several recent studies, largely using immunohistochemical procedures, claim to have identified other potentially important peptides in the posterior pituitary, including met-enkephalin, leu-enkephalin, dynorphin, somatostatin and gastrin. However, one cannot but question some of these studies, for it is quite possible that the antisera used may have cross-reacted with parts of the oxytocin or vasopressin prohormone, or with other cellular proteins. On the other hand, the magnocellular neurones certainly have the genetic information with which to construct the peptides in question, and the genes for the production of these peptides may not be totally suppressed. A very interesting new range of questions arises if in fact these neurones really do produce these regulatory peptides. Are they packaged within the oxytocin- and vasopressin-containing granules? Are they released concurrently with the major hormones? What is their function after release?

Most research on axon transport in magnocellular neurones has concentrated on the movement of materials in the orthograde direction, namely from the cell body to the nerve terminal. Neurones do, however, transport materials in the reverse or retrograde direction, and there is the distinct possibility that materials may move in both directions at the same time. The physiological significance of retrograde transport is largely unknown but it could, for example, serve to return granule membrane to the cell body after the contents have been released. To the neuro-anatomist retrograde transport is a useful tool through which to trace nerve pathways. Horse-radish peroxidase, or HRP as it is commonly abbreviated, is readily taken up by nerve endings and is then transported in a retrograde direction. Thus HRP placed in the posterior pituitary finds its way back to the cell bodies of the magnocellular neurones that innervate the pituitary.

Storage and turnover of neurosecretory granules
The posterior pituitary contains an enormous store of oxytocin and vasopressin when compared with the amounts required to elicit normal physiological functions. The pituitary of the rat, for example, contains about 500 mU ($= 1 \mu$g) oxytocin and 700 mU vasopressin; as little as 500 μU (1 ng) intravenously will evoke a milk ejection, and the kidney is even more sensitive to the antidiuretic actions of vasopressin. These large stores of hormone could be of importance under emergency conditions, during the expulsive stage of labour or after haemorrhage, for example. Conversely, it is important to appreciate that peptide synthesis does not

have to be linked to the rate of release other than over a relatively long time scale (i.e. days).

John Morris, whilst working in the Bristol group, estimated the number of neurosecretory granules in the posterior pituitary of the rat to be 2×10^{10}. On the basis of this score each granule must contain, if most of the hormone is stored within the granules, some 60 000 molecules of either oxytocin or vasopressin and their associated precursor peptides. Only about 30 per cent of these granules are found in terminals that abut onto capillaries; the majority are in the non-terminal dilatations (Herring bodies) (Fig. 2.8). This provides an anatomical basis for the concept of a readily releasable pool of peptide, though in reality there is no sharp division between what is releasable and what is not. Estimates of the readily releasable pool suggest that it may represent 10 per cent of the gland content, but this is still a large amount of hormone when one remembers that milk ejection requires the release of only 0.1 per cent of the pituitary's store of oxytocin. Curiously it is the newly synthesized hormone that is preferentially released, which suggests that new neurosecretory granules entering the posterior pituitary must pass first to the terminals before moving on to less available storage sites (Fig. 2.8).

A daily turnover rate of about 20 mU oxytocin, which is 4 per cent of total content, has been estimated for a sexually inactive male rat in water balance. What such turnover constitutes remains very much a matter for speculation: it could represent basal release due to low levels of electrical activity or it could relate to breakdown *in situ*.

Fig. 2.8. Movement and storage of neurosecretory granules in the axon dilatations and capillary-based terminals of magnocellular neurones within the posterior pituitary. The relative distribution of granules between the axons, the Herring bodies and the terminal dilatations has been measured by quantitative electronmicroscopy, and the movement of the granules determined by autoradiography. (Data from J. F. Morris. *Rec. Prog. Horm. Res.* **31**, 243–94 (1975).)

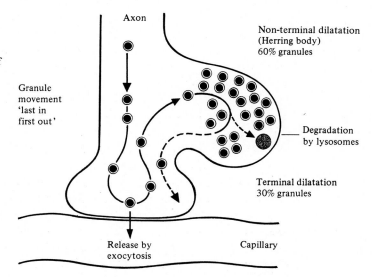

Axon

Non-terminal dilatation (Herring body) 60% granules

Granule movement 'last in first out'

Degradation by lysosomes

Terminal dilatation 30% granules

Release by exocytosis

Capillary

Electrophysiological properties of magnocellular neurones

The neurones of the magnocellular nuclei generate action potentials, and these potentials resemble in virtually every respect the potentials generated by neurones elsewhere in the nervous system. Through the release of their various neurotransmitters, the 5000 or more synapses impinging on each neurone cause either depolarization (= excitation) or hyperpolarization (= inhibition) of the post-synaptic cell membrane. When depolarization reaches a critical threshold, the cell body generates a self-propagating action potential or nerve impulse that sweeps down the axon to invade the terminals of the posterior pituitary. Thus the incidence (or frequency) with which action potentials are generated is a function of the intensity and balance of these numerous excitatory and inhibitory synaptic inputs. The individual synaptic events are not easy to record and require an electrode to be placed *within* the cell, but the action potentials present no problem and can be monitored with a microelectrode, usually a glass micropipette of 1 μm tip diameter filled with 3M NaCl, stereotaxically placed to within 30 μm of the cell body (Fig. 2.9).

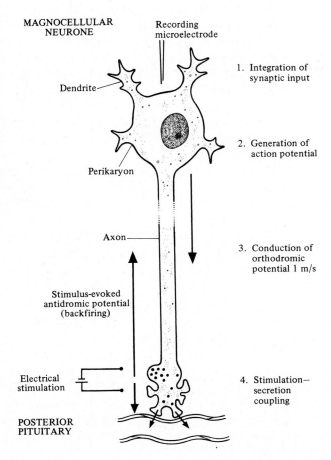

Fig. 2.9. The magnocellular neurone of Fig. 2.4 redrawn to illustrate its electrophysiological properties. The stages in the production of the action potential, its conduction to the posterior pituitary and the induction of hormone release are given on the right-hand side. Electrical stimulation of the posterior pituitary with an implanted electrode causes hormone release and the conduction of action potentials back to the cell body, in what is termed the antidromic direction.

MAGNOCELLULAR NEURONE

Recording microelectrode

Dendrite

Perikaryon

Axon

Stimulus-evoked antidromic potential (backfiring)

Electrical stimulation

POSTERIOR PITUITARY

1. Integration of synaptic input

2. Generation of action potential

3. Conduction of orthodromic potential 1 m/s

4. Stimulation–secretion coupling

There are currently three main approaches to the measurement of electrical activity. The now almost traditional approach involves an acute experiment on an anaesthetized animal restrained within a stereotaxic frame. This has the advantage of permitting the monitoring of many physiological variables in parallel with electrical recordings taken from one or more sites, but is invariably handicapped by the fact that anaesthetics tend to disrupt the processes one wishes to study. The situation is changing, however, and new anaesthetics designed to eliminate pain and stress without compromising the neuroendocrine situation under study promise to give this approach new potential. An alternative is to remove the brain quickly and prepare a 100 μm slice of the hypothalamus for culture in a tissue bath. These slices remain viable for up to 24 h, facilitating intracellular recordings, and enabling the response of neurones to putative neurotransmitters and regulatory peptides ejected into their local environment (iontophoresis) to be examined more easily than in the intact animal. The size of the animal is also less of a problem: one could work equally well with a slice of tissue from a cow or a mouse. The third approach is to record from the conscious unrestrained animal; not impossible but difficult. Alastair Summerlee, working in our laboratory, has pioneered a technique involving the chronic implantation of microwire electrodes, and using this approach has obtained recordings of action potentials from single magnocellular neurones for periods of several days during both labour and lactation in the conscious rat. Two problems still remain: it is difficult to identify the cells from which the recordings are obtained and to monitor other physiological variables in parallel.

Using any of these technical approaches one finds that a proportion of magnocellular neurones exhibit spontaneous activity: the neurones continue to generate action potentials in the absence of any obvious sensory input. The *mean* rate of this discharge is usually in the range of two to five action potentials per second, but the distribution is exponentially arranged with most neurones firing at less than 1/s. As a consequence the mean rate of activity that is recorded is very much a function of the experimenter's skill in detecting neurones with little or no spontaneous activity before they are destroyed by the advancing microelectrode. It is therefore difficult to make comparisons of firing rates between one study and another.

Antidromic identification.

Action potentials can be made to travel in reverse direction from the posterior pituitary to the magnocellular nuclei (antidromically) (Fig. 2.9). This does not occur in nature as far as we are aware, but as an experimental tool it has revolutionized the electrophysiological study of magnocellular neurones since it provides an on-line method of selectively identifying those hypothalamic neurones that send axons into the posterior pituitary. Such identification is achieved by first inserting a bipolar stimulating electrode of about 500 μm diameter into the posterior pituitary. The position of this electrode can be verified since, if it is correctly placed, the application of a 5 s train of electrical stimuli should release sufficient oxytocin to produce a rise in intramammary pressure in a lactating animal (see Fig. 2.12). At the same time each stimulus pulse applied to the pituitary will cause a single action potential to pass antidromically along individual axons towards the cell body. Antidromic potentials recorded from the cell body very closely resemble orthodromic potentials generated by synaptic activation. Synaptic relays could exist between the point of stimulation and recording, but these we can confidently exclude when recordings satisfy three criteria: the antidromic potentials should recur at a constant latency after stimulation, should follow on a 1:1 basis stimulus pulses delivered at high frequency (50–200/s), and should exhibit 'collision'. This last test is achieved by arranging for an orthodromic potential – recorded from the cell body as

Fig. 2.10. Extracellular recordings from an oscilloscope of action potentials obtained from single neurones in the magnocellular nuclei of the rat with a stereotaxically placed microelectrode. Part (*a*) illustrates, on a relatively fast time scale, antidromic potentials evoked by electrical stimulation of the posterior pituitary; the stimulation artifacts that occurred at the time of stimulation have been erased to avoid confusion. In *a*(1) two antidromic potentials have been produced by two stimulation pulses after a latency of about 12 ms: this latency is a product of the conduction velocity and the length of the axon. In *a*(2) an orthodromic potential generated synaptically at the cell body has been used to fire the stimulator. Thus the first antidromic potential passing up the axon meets the orthodromic potential passing down, and the conduction of both is terminated. Part (*b*) illustrates, on a time scale that is 100 times slower, the 'spontaneous' generation of a random train of orthodromic potentials by a putative oxytocin-producing neurone.

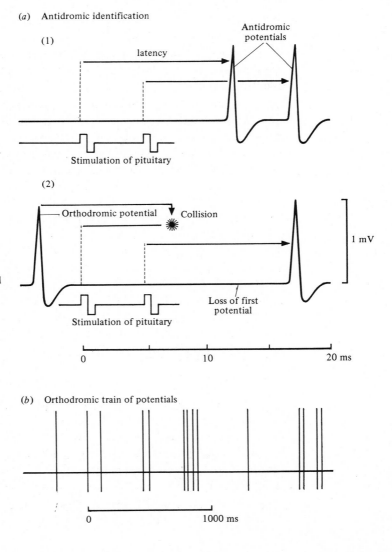

(*a*) Antidromic identification

(1)

latency

Antidromic potentials

Stimulation of pituitary

(2)

Orthodromic potential   Collision

1 mV

Loss of first potential

Stimulation of pituitary

0   10   20 ms

(*b*) Orthodromic train of potentials

0   1000 ms

it sets out on its journey – to itself trigger the immediate stimulation of the posterior pituitary. In this way the orthodromic potential passing down should collide midway along the axon with the antidromic potential passing up, thus cancelling further conduction of the antidromic potential (Fig. 2.10).

The latency between the application of the stimulus to the pituitary and the recording of the antidromic potential in the hypothalamus depends on the length of the axon and its conduction velocity. This latency is 7–15 ms in the rat and 15–25 ms in the rabbit, which indicates a conduction velocity of less than 1 m/s. The 'slow' conduction of the action potential, relative to the 100 m/s of nerve impulses in large myelinated axons, should not be confused with the 'fast' transport of the neurosecretory granules. Action potentials still move 10 000 times faster than granules along the axons of these magnocellular neurones.

Stimulus–secretion coupling

The invasion of the axon terminal by the action potential results in a transitory influx of sodium ions, and this in turn leads to the opening of calcium channels so that calcium pours into the terminal (Fig. 2.11). As a consequence of this increase in intracellular calcium, neurosecretory granules move towards the outer membrane of the terminal, the granule and terminal membranes fuse and the granule contents are discharged (exocytosis). From here the peptide diffuses into the capillaries and within a few seconds appears in the vascular circulation. The problem now is to

Fig. 2.11. Stages in stimulus–secretion coupling within an axon terminal of the posterior pituitary.

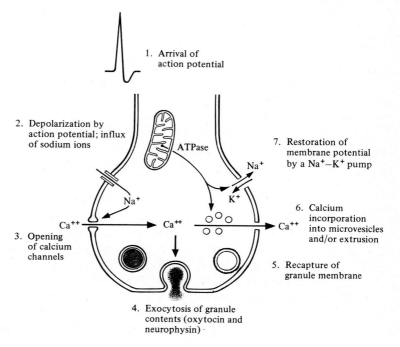

1. Arrival of action potential

2. Depolarization by action potential; influx of sodium ions

ATPase

7. Restoration of membrane potential by a Na^+–K^+ pump

Na^+

Na^+

K^+

6. Calcium incorporation into microvesicles and/or extrusion

Ca^{++} → Ca^{++} → Ca^{++}

3. Opening of calcium channels

5. Recapture of granule membrane

4. Exocytosis of granule contents (oxytocin and neurophysin)

restore the *status quo* within the terminal. Something has to be done about the increase in terminal membrane, the intracellular levels of calcium have to be reduced, and the membrane potential restored. Studies in which the granule membranes have been labelled with [^{3}H] choline suggest that granule membrane is recaptured and appears in the terminals as large vacuolated vesicles before transport back to the cell body. Recent evidence obtained independently by John Morris and Jean Nordmann and their colleagues has rather changed our views on what happens to the calcium; they suggest that excess calcium is sequestered into microvesicles before being removed from the terminal. The membrane potential is restored as in other neural tissue by an energy-dependent sodium–potassium interchange.

Considerable controversy did exist as to the validity of the process of exocytosis just described, and the problem was fuelled by the failure of electronmicroscopists to produce convincing illustrations. In retrospect, one should have been more surprised by their ability to produce evidence of it at all. Exocytosis is an extremely fast process, perhaps taking less than 10 ms. Therefore, even when one artificially elevates the level of release to say 1 per cent of the gland content per minute, one would only expect to capture 1–2 granules/million in the process of exocytosis. Dennise Theodosis and Jean Jacques Dreifuss working with freeze-fracture techniques in Geneva have now resolved the problem and produced abundant evidence (Fig. 2.5). The great advantage of the freeze-fracture technique is that it allows large areas of the terminal membrane to be scanned, compared with the cross-sections provided by traditional electronmicroscopy.

Frequency facilitation. From the foregoing evidence one is given to think that hormone release relates directly to the number of action potentials entering the nerve terminals of the posterior pituitary. In fact this greatly underrates the ability of the system. Action potentials entering the terminals at a rate of 50/s release 100–1000 times more oxytocin, *per potential*, than those arriving at the rate of say 2/s. This enormous frequency facilitation is probably caused by the calcium influx evoked by one potential summating with the calcium entering with the next potential. Fig. 2.12 illustrates the effects on milk ejection of applying different frequencies of stimulation to the posterior pituitary. Such an analysis combines the dynamics of hormone release with the response characteristics of the mammary gland. The optimal parameter of stimulation for the rat, in terms of the contribution *each* stimulus pulse makes towards the amplitude of the mammary contraction, falls between a pulse train of 1.3 seconds of stimulation at 67 pulses/s and a train of 3.3 seconds at 40 pulses/s. Similar parameters of stimulation have also been found to be optimal for the induction of hormone release from the posterior pituitary maintained *in vitro*. It is so interesting to think back a moment now we know something of this process of frequency facilitation. When the late

Geoffrey Harris first stimulated the posterior pituitary and hypothalamus he had to be content with the 50-cycle frequency of the mains electricity supply: the Electricity Board could not have provided a better frequency for the release of oxytocin!

The physiological significance of frequency facilitation cannot be stressed too highly. On the one hand, if cells are made to fire at high frequencies they become very efficient at releasing hormone, and herein probably resides the basis for pulsatile hormone secretion. Conversely, low levels of spontaneous activity, which for the moment we may consider as 'noise', will not cause a proportional drain on the peptide stores.

Opioid inhibition of the peptide terminal. Frequency facilitation may have made the life of the electrophysiologist more difficult, but in 1979 we stumbled on a new phenomenon that makes the association of electrical activity and hormone release yet more complex. Some anaesthetized lactating rats on which we were working were observed not to release oxytocin in response to electrical stimulation of the posterior pituitary until given the highly selective opiate antagonist, naloxone. Immediately, therefore, we had evidence to suggest that endogenous opioid peptides were inhibiting oxytocin release at the level of the posterior pituitary. Subsequent experiments in our laboratory and elsewhere, both *in vivo* and *in vitro*, have shown that opiates, like morphine, and opioid peptides, such as β-endorphin

Fig. 2.12. Effect of electrical stimulation of the posterior pituitary on milk ejection in the lactating rat, measured from the rise in pressure recorded from a cannulated galactophore. One millisecond biphasic square-wave pulses were applied at a constant peak-to-peak current of 300–500 μA. In the upper three situations the number of pulses applied has been kept constant, whilst the frequency and period of stimulation have been varied. In the lower three examples, the period of stimulation has been kept constant. On the right, an approximate 'index of efficiency' has been determined by dividing the amplitude of the mammary contraction by the number of stimulation pulses applied to the pituitary. The optimal frequency of stimulation is 40–50 Hz.

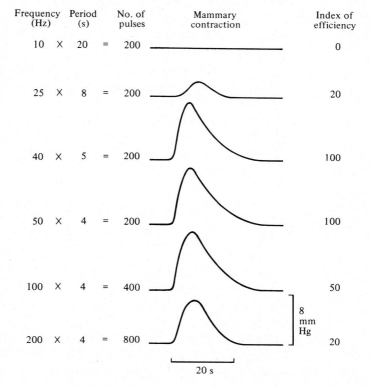

Frequency (Hz)		Period (s)		No. of pulses	Mammary contraction	Index of efficiency
10	×	20	=	200		0
25	×	8	=	200		20
40	×	5	=	200		100
50	×	4	=	200		100
100	×	4	=	400		50
200	×	4	=	800		20

8 mm Hg

20 s

and enkephalin analogues, reduce the amount of oxytocin and vasopressin released by electrical stimulation. There is some evidence to suggest that this dissociation of stimulus–secretion coupling may be of physiological importance. The posterior pituitary contains the first requirement, namely, high affinity membrane-bound opiate receptor sites. Secondly, the neuro-intermediate lobe of the pituitary releases β-endorphin during stress and there is every reason to think that this peptide must reach the terminals of the magnocellular neurones in the adjacent posterior pituitary, because they are outside the blood–brain barrier. Evidence is also accumulating from studies on other parts of the nervous system to indicate that this opioid inhibition of the nerve terminal may be a fairly widespread phenomenon, and the generally accepted view is that it may act through an immobilization of calcium. Clearly this is of immense importance in reproductive biology. Opioid peptides could, for example, inhibit the release of the hypothalamic hormones (GnRH) that regulate the anterior pituitary gland, by an action on the nerve terminals that impinge on the portal blood vessels of the median eminence (see Chapter 1); furthermore, in a clinical context, it means that the administration of analgesics, such as pethidine, to women in labour may compromise their ability to release their own oxytocin.

There is, however, the possibility that we may have an even more subtle phenomenon at hand. Evidence has been presented within the past year to suggest that enkephalins are present in oxytocin-producing neurones, albeit in small amounts, and may therefore be released in parallel with the hormone. If this were to occur, the enkephalin might then act back onto the terminal from which it has just been released. Whilst I am admitting to a spot of speculation, such direct feedback at the peptide-secreting nerve terminal could serve to limit the duration of secretion during pulsatile discharge.

Stimulus–secretion coupling is therefore more complex than we had thought hitherto: no longer can we assume a direct relationship between electrical activity recorded from the cell body and peptide release at the nerve terminals.

Development of a model for the study of electrical activity during hormone release

Electrical stimulation of the posterior pituitary provides only a simulation of the events that may occur during the process of secretion: proof of what actually happens demands that one monitors electrical activity during the physiological release of the hormone. This is easier said than done. Ideally, one needs a situation where one can readily manipulate the oxytocin neurones with a physiological stimulus, and simultaneously measure hormone release second by second. There are two situations where this may be possible. These involve the study of electrical activity during oxytocin release evoked by the sucking of the nipples (the milk ejection reflex) and dilatation of the reproductive tract (the Ferguson reflex).

The afferent side of these two reflexes, i.e. the sensory input to the brain, is subject to many influences. Inhibitory pathways within the brain, activated by pain, anxiety or stress, readily block the central transmission of the sensory stimuli one wishes to study, though we are progressively learning to counteract these by the administration of selective neuro-transmitter antagonists. There is on the other hand evidence to suggest that steroid hormones may 'prime' the sensory input. In women, for example, the sensitivity of the nipple, the surrounding areolar tissue and even the rest of the breast, changes greatly according to reproductive state, and in particular a very substantial increase in sensitivity occurs in the hours following childbirth (Fig. 2.13). It has also been shown that breast stimulation in late pregnancy will not produce an oxytocin discharge, whereas it will immediately after delivery. Therefore, one cannot afford to assume that oxytocin release will occur in response to what may appear to be an *appropriate* stimulus. When it works, the sucking of the nipples is an excellent stimulus because it is selective and releases oxytocin without vasopressin.

Oxytocin is not an easy hormone to measure by the common procedure of plasma collection and radioimmunoassay, and this approach cannot provide the temporal resolution required for the evaluation of electrical events that may only last for seconds. For once, the answer lies in the on-line measurement of a biological response. There are two target tissues that one can use for this purpose – the mammary gland and the uterus – but they differ substantially in their response characteristics. The oxytocin receptors

Fig. 2.13. Changes in the tactile sensitivity of the cutaneous tissue of the human breast at puberty and during the onset of lactation. Sensitivity was scored as the minimum distance that subjects could separately discriminate when gentle pressure was applied through the points of a pair of dividers. The change in the sensitivity of the areolar region was much greater than that shown above, but this was difficult to quantify because during pregnancy subjects could not discriminate two points of stimulation applied across the full diameter of the areolar region. (Data from J. E. Robinson and R. V. Short. *Brit. Med. J.* **1**, 1188–91 (1977).)

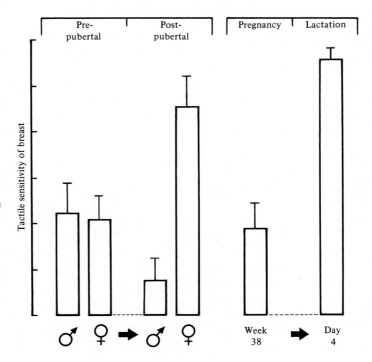

of the uterus only develop within a few hours of parturition and disappear almost as quickly (Fig. 2.14); this means that the sensitivity of the uterus to oxytocin, as measured in the amplitude or frequency of contractions, changes dramatically over a short period of time. Secondly, the myometrial cells of the uterus are coupled so that a bolus injection of oxytocin will tend to evoke a series of contractions that may well outlast the appearance of the hormone in circulation, given that oxytocin has a half-life of 1–2 minutes. The mammary gland does not display such marked changes in receptor concentration (Fig. 2.14) and bolus injections of oxytocin, within a given range at least, cause only single contractions, lasting about 15 seconds in a rat but longer in larger species. What an advance it would represent to have a probe that we could insert into the bloodstream for the second-by-second measurement of the peptides flowing past!

The best approach in the past, it seemed, was to record the electrical activity of magnocellular neurones in the lactating rabbit, apply the young to the nipples and monitor oxytocin release through the measurement of intramammary pressure. Unfortunately, the anaesthetics that had to be administered to allow electrophysiological recordings to be made abolished the milk ejection reflex of the rabbit, and there progress was halted throughout the 1960s. No one in this period seems seriously to have considered the rat as an alternative. Perhaps we were too conditioned by the classical studies on milk ejection in the rabbit by Geoffrey Harris and Barry Cross to be able to think of milk ejection in any other terms. The rabbit displays such a profound milk ejection reflex: the doe suckles for precisely 3–4 minutes once a day, and the young imbibe in this brief period up to 20 per cent of their body weight in milk. Clearly the rat does not

Fig. 2.14. Concentration of oxytocin receptors in the myometrium and mammary gland of the rat during pregnancy and lactation. The concentration of receptors is expressed as the amount of specific binding of tritiated oxytocin to particulate fractions (femtomoles/mg protein). (Data from M. S. Soloff, M. Alexandrova and M. J. Fernstrom. *Science*, **204**, 1313–15 (1979).)

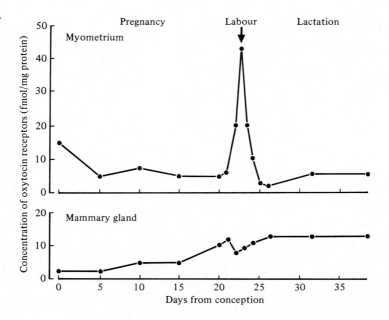

nurse so infrequently or milk eject so dramatically, but had one paused to consider that a rat spent up to 18 h each day with her young attached to the nipples one might have questioned how milk ejection was organized. Quite by chance, Jonathan Wakerley, whilst conducting studies we had designed to explore the Ferguson reflex in the anaesthetized rat in the autumn of 1970, observed that pups he had placed on the nipples simultaneously stood to attention every 5–10 minutes and sucked with increased vigour for 15–20 seconds. Within days the experimental situation was duplicated and measurements made of intramammary pressure. Each behavioural response of the pups was observed to *follow* within a fraction of a second an abrupt rise in intramammary pressure (Fig. 2.15). Further, an intravenous bolus injection of 300–1000 μU oxytocin evoked, after a latency of about 10 s, an identical pressure rise and behavioural response. Reflex milk ejection in the rat was intermittent despite the continuous attachment of the young to the nipples.

It was some years before we understood our good fortune. Rats only milk eject when they are asleep, and the reason we had observed milk ejection under anaesthesia was due to the fact that most anaesthetics induce a pattern of brain activity akin to slow-wave sleep. Conversely, any stimulus that prevents the unanaesthetized rat from sleeping blocks milk ejection; the presence of an observer is usually sufficient, and that probably accounts for why the dramatic behavioural response of the young to the rise in pressure at milk ejection had not been noted previously under normal husbandry conditions. A word of caution is perhaps appropriate for anyone entering the field; rats can sleep with their eyes open and remain awake when their eyes are shut.

Fig. 2.15. A recording of intramammary pressure to illustrate reflex milk ejection in the anaesthetized rat. The mother was separated from her young overnight. The next morning she was anaesthetized with 6 mg/kg xylazine and 6 mg/kg diazepam, both given intramuscularly, and the main galactophore of one of the inguinal mammary glands was cannulated for the measurement of intramammary pressure, with the results shown. Eleven pups were applied to the nipples and left there throughout the period shown by the event bar. Individual milk ejections commenced within a few minutes of applying the young to the nipples and recurred at regular intervals until the young were removed. A milk ejection recorded on a much faster time scale is shown lower right.

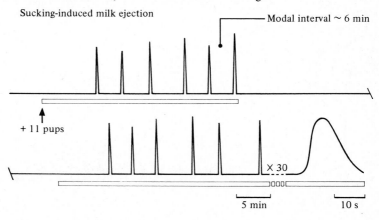

Sucking-induced milk ejection

Modal interval ~ 6 min

+ 11 pups

× 30

5 min   10 s

The milk ejection reflex of the rat makes an excellent laboratory demonstration that cannot fail to fascinate the audience. Take a rat of about 330 g at Day 7–9 of lactation rearing 10–12 strong pups and separate her from all but one of her young overnight. Next day anaesthetize the doe with 35 mg/kg pentobarbitone sodium, place her supine, and 1 h later apply 10 or more of the now hungry babies to the nipples. After 15–30 min of sucking and at intervals of 3–10 min thereafter the young will synchronously rise to their feet and suck with all their might. Many other anaesthetics may be used with equal success, but one is advised to avoid those that contain opiates for the reasons discussed earlier.

The anaesthetized lactating rat thus provided the model that we sought for the electrophysiological analysis of the events underlying the pulsatile release of a peptide hormone.

Fig. 2.16. A recording of electrical activity from an antidromically identified paraventricular neurone in an anaesthetized rat during oxytocin release induced by the sucking of the young. Part (*a*) illustrates the explosive increase in the number of action potentials generated by these putative oxytocin-producing neurones about 10 s before milk ejection. Each deflection corresponds to a single action potential; the data are also shown in a histogram. (*b*) is a simultaneous recording of intramammary pressure to indicate the time of milk ejection, and (*c*) shows radar recordings of the increased activity of the pups at milk ejection and in the period immediately following. Note, no increase in sucking activity is observed coincident with the activation of the magnocellular neurones; at this time the pups, whilst firmly attached to the nipples, remain quite somnolent.

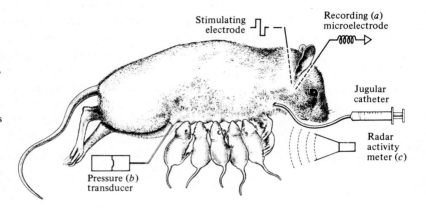

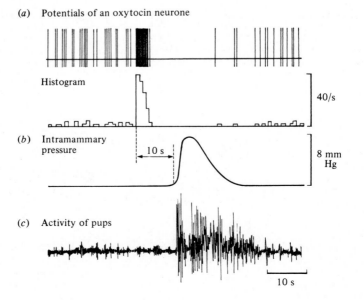

(*a*) Potentials of an oxytocin neurone

Histogram

40/s

(*b*) Intramammary pressure

10 s

8 mm Hg

(*c*) Activity of pups

10 s

Electrical activity related to the pulsatile release of oxytocin

Approximately 50 per cent of antidromically identified magnocellular neurones in the supraoptic and paraventricular nuclei of the anaesthetized rat were subsequently found to display an explosive burst of electrical activity some 10–12 s before the rise in intramammary pressure at milk ejection (Fig. 2.16). Activity increased from 0–4 action potentials/s to 30–80 action potentials/s for 2–4 seconds; thereafter there was usually a 20–40 s period of quiescence. The number of neurones displaying these bursts of activity was not increased when the number of babies attached to the nipples was raised from 7 to 11, nor was the interval between one burst and the next reduced, though the number of action potentials within each burst and the amount of oxytocin released was somewhat increased. When the number of young was reduced to six or less, under anaesthetized conditions, bursts of electrical activity and milk ejection ceased. The latency observed between activation and milk ejection corresponds with that recorded after electrical stimulation of the posterior pituitary, and may be broken down approximately as follows: 2 s for release into the circulation, 4 s for the hormone to circulate to the mammary gland, and 6 s for the activation of the myoepithelial cells. Most important, the peak frequency of discharge capitalizes fully on the process of frequency

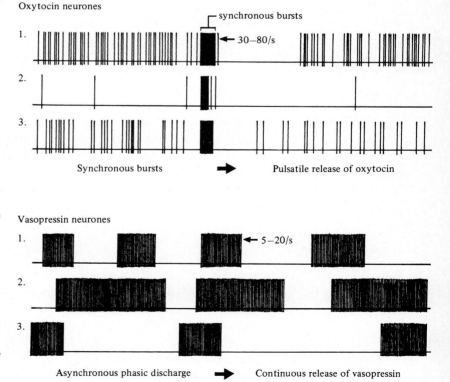

Fig. 2.17. Recordings of action potentials from three putative oxytocin-producing and three vasopressin-producing neurones in the rat to illustrate the contrasting patterns of electrical activity observed during the sucking of the young. The oxytocin-producing neurones display synchronized bursts of activity (30–80 action potentials/s) about 10 s before milk ejection, despite the continuous attachment of the young to the nipples. This induces the release of an oxytocin pulse of optimal size for the induction of a mammary contraction. By contrast, many vasopressin-producing neurones fire in asynchronous phases unrelated to milk ejection. Such an asynchronous pattern of electrical activity probably generates a continuous low level of vasopressin release.

facilitation discussed earlier. Within the past two years we have extended these studies to the unanaesthetized rat, and have observed virtually identical bursts of electrical activity immediately before milk ejection.

As far as can be judged, these explosive bursts of activity occur in the same 2–4 second time period in all four magnocellular nuclei (Fig. 2.17). The effect of such synchronous activation of the 9000 or so oxytocin-producing cells in the rat hypothalamus is that about 500 000 action potentials pass to the posterior pituitary within a period of 2–4 seconds. In very approximate terms, each action potential leaving the magnocellular nuclei during this period of accelerated activity releases 3 fg (femtograms, 10^{-15}g) oxytocin or the contents of 40 neurosecretory granules.

Putative vasopressin-producing neurones rarely display any change in electrical activity during suckling, which is rather amazing when one considers the magnitude of the activation intermittently displayed by oxytocin cells situated in some cases no more than 50 μm distant. In their own way, however, the patterns of electrical activity displayed by vasopressin-producing neurones are just as remarkable. In response to haemorrhage, dehydration or increases in plasma osmotic pressure, the vasopressin neurones of the rat accelerate their activity and then enter a curious mode of discharge in which periods of relatively brisk firing (5–20 action potentials/s) alternate with periods of silence (Fig. 2.17). Such periods of phasic discharge, as we have termed the condition, are not synchronized between adjacent neurones, which suggests that the mechanism determining this pattern of discharge resides within the individual neurone. These phasically firing neurones could represent the actual osmoreceptors of the brain, as proposed by Verney in the 1920s, but this is still a matter of much debate.

The main reason for including this discussion of the vasopressin neurone is that it illustrates for reproductive biologists an alternative strategy for the regulation of hormone release. Richard Dyball, now at King's College, London, has shown that phasic activity releases more vasopressin than would occur if the same number of potentials were equally spaced; presumably the higher rate of firing during each phase of activity generates a degree of frequency facilitation. It is not possible to measure the release of vasopressin second by second, but one would predict that a pattern of continuous hormone release would result from the asynchronous firing of the neurones, with the actual level of release being set by the relative lengths of the active and silent phases, and the frequency of discharge. The final point has not to my knowledge been made before, but clearly the genome for oxytocin and vasopressin biosynthesis has to be linked to the genome determining the respective patterns of electrical activity.

Central regulation of oxytocin release

Nothing is known of how the explosive bursts of electrical activity displayed by the oxytocin cells are generated, or what determines the interval between one burst of activity and the next. Activation depends on the sucking of the young, but there is no progressive increase in the firing of the neurones as the time of activation approaches and there is no increase in firing at the time of milk ejection or in response to milk ejection evoked by exogenous oxytocin, despite the vigorous increase in sucking that occurs then. The interval between one milk ejection and the next is relatively regular, and the interval cannot be reset by introducing additional milk ejections with exogenous oxytocin or electrical stimulation to the posterior pituitary. Collectively these observations suggest the presence within the nervous system of a pulse generator which is activated by the sucking stimulus and which in turn instructs the magnocellular neurones to discharge synchronously.

A considerable amount is known concerning the neurotransmitters involved in the nervous pathways leading to the activation or inhibition of oxytocin release. This knowledge has largely developed from studies in which neurotransmitter agonists, antagonists and synthesis inhibitors have been administered into the cerebral ventricles of the rat brain and their effects on reflex milk ejection determined. A model presenting some of these interactions is shown in Fig. 2.18. Two points only will be discussed to illustrate the potential of such a pharmacological dissection. The sucking-induced release of oxytocin in the rat is very effectively blocked by the

Fig. 2.18. A model depicting the neurotransmitter mechanisms involved in the reflex milk ejection of the rat. We suggest from extensive studies with selective neurotransmitter antagonists that cholinergic (Ach), dopaminergic (D) and noradrenergic (NA) neurones are implicated, and that the cholinergic and noradrenergic relays involve nicotinic and α-type receptors, respectively. The cholinergic input could in part be the pre-synaptic facilitation of a dopaminergic system that may simply serve to generate the periodicity between one milk ejection and the next. The reflex is also inhibited (−) by noradrenaline, opioid peptides and adrenaline acting on the central pathway, the posterior pituitary and the mammary gland, respectively. (For further details see E. Tribollet, G. Clarke, J. J. Dreifuss and D. W. Lincoln. *Brain Res.* **142**, 69–84 (1978) and G. Clarke, C. H. D. Fall, D. W. Lincoln and L. P. Merrick. *Br. J. Pharmac.* **63**, 519–27 (1978).)

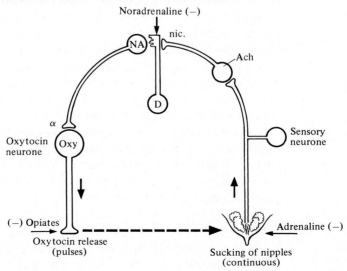

Noradrenaline (−)

nic.

NA

Ach

D

α

Oxytocin
neurone
Oxy

Sensory
neurone

(−) Opiates

Adrenaline (−)

Oxytocin release
(pulses)

Sucking of nipples
(continuous)

central administration of antagonists to acetylcholine of the nicotinic type, such as mecamylamine, but not by those of the muscarinic type, like atropine, which suggests the involvement of a cholinergic (nicotinic) synapse in the sensory pathway from the nipples to the magnocellular nuclei. Carbachol, a drug that mimics the actions of acetylcholine, readily evokes the release of oxytocin when placed in the cerebral ventricles and causes a profound and sustained activation of the magnocellular neurones. Surprisingly, this effect is blocked by atropine and not by mecamylamine, and that points to the existence of a second oxytocin-releasing pathway involving a cholinoceptor (muscarinic) synapse. The sight, sound or thought of a baby crying is often sufficient to trigger milk ejection in a lactating woman, and music or the rattling of the food bucket may work in the cow byre, so connections must exist between higher nervous centres and the magnocellular nuclei; it could be these that are activated by carbachol.

Reflex milk ejection in the rat is also blocked in a dose-dependent manner by the central administration of adrenoreceptor antagonists of the alpha type, such as phentolamine, but not the beta type, such as propranolol; thus we may deduce that an α-adrenoreceptor mediates at some point the sucking-induced release of oxytocin. But the situation is more complex; some anaesthetized rats do not milk eject in response to the sucking of the young unless given a centrally active β-adrenoreceptor antagonist, and this points to the existence within the brain of an inhibitory pathway involving a β-adrenoreceptor synapse. Central inhibition of oxytocin release is indeed a well documented phenomenon, and is illustrated by the fact that many women cannot milk eject when required to nurse in strange company. The endogenous ligand for both these receptors is probably neuronally released noradrenaline, and this creates an interesting situation. Noradrenaline may be involved in the sucking-induced reflex via alpha receptors, and in the inhibition of the reflex via beta receptors. This is, of course, made possible by the fact that neurotransmitters function within the immediate vicinity of the synapse and do not circulate within the brain. Such observations also highlight the futility of administering noradrenaline or other neurotransmitters, rather than selective antagonists.

Background electrical activity and the basal release of oxytocin
Having seen how synchronized high frequency bursts of electrical activity may be used to fashion the pulsatile release of oxytocin, we may now return to the question of what significance the background electrical activity of the cells may have on the basal release of hormone. Most oxytocin cells discharge continuously (day and night) at a mean rate of 2–4 action potentials/s; this is not a fast rate but it is sustained, and it could amount to the 9000 oxytocin cells of the rat generating more than 2000 million action potentials per day. If each of these potentials released the 3 fg of oxytocin calculated during high frequency firing, the pituitary

would be emptied of its vast store of oxytocin in just 1–2 hours. This does not occur, and biochemical studies suggest that the daily loss of oxytocin from the posterior pituitary in a non-stimulated rat may be as low as 4 per cent. One should not on the other hand jump to the opposite conclusion and adopt the view that background electrical activity releases no hormone. If each potential released just 3 attograms $(3 \times 10^{-18}\text{g})$ of oxytocin, the basal level of oxytocin in the plasma would be in the low picogram/ml range, as indeed is the case. Moreover small changes in background discharge, of as little as one action potential per second, would cause, because of its sustained nature, a significant change in the daily turnover of oxytocin and of the basal level of oxytocin in the plasma.

Back in the 1960s the late Hans Heller reported changes of up to 50 per cent in the oxytocin content of the posterior pituitary during the oestrous cycle of the rat, with the content falling steeply between oestrus and metoestrus. These changes appear excessive and merit reinvestigation with current assay methods. There is on the other hand evidence to indicate that the spontaneous activity of magnocellular neurones changes during the oestrous cycle of the rat, being highest at pro-oestrus and oestrus, and that after ovariectomy it is elevated by oestrogen treatment (Fig. 2.19). The physiological significance of the basal levels of oxytocin in plasma and of the changes that may occur is by no means clear, though in the past year two studies have been published in which such basal levels of oxytocin could be implicated. Whilst I was working in Perth with Marilyn Renfree, we obtained evidence to the effect that basal levels of oxytocin might evoke milk ejection in the one mammary gland of the agile wallaby to which the pouch young was attached, without affecting the animal's other lactating mammary gland, which was some 200 days further into lactation and feeding the juvenile now running at foot. The second piece of evidence comes from the work of John McCracken working at the Worcester

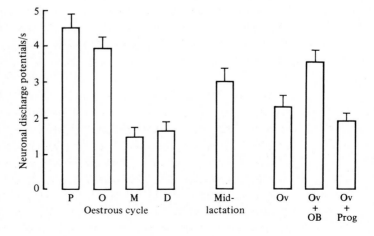

Fig. 2.19. Changes in the number of action potentials generated ('spontaneous activity') by magnocellular neurones in the rat during the oestrous cycle, lactation and following ovariectomy (Ov), with oestrogen (OB) and progesterone (Prog) treatment. P: pro-oestrus; O: oestrus; M: metoestrus; D: dioestrus. These data suggest an oestrogen-dependent increase in the background electrical activity of the magnocellular nuclei. (Data from H. Negoro, S. Visessuwan and R. C. Holland. *J. Endocr.* **59**, 545–58 (1973).)

Foundation for Experimental Biology. He has produced compelling evidence to indicate that ovarian steroid hormones (oestrogen and pro- gesterone) regulate the production of oxytocin receptors in the uterus of the ewe. Oxytocin then acts on these receptors to evoke the release of the luteolytic factor, prostaglandin $F_{2\alpha}$, and therein could reside the control of the oestrous cycle (see Chapter 5). Of course, in both circumstances these oxytocin-mediated effects might be enhanced if steroid hormones also increased the level of oxytocin secretion, but that may not be the whole story. Clare Wathes and Ray Swann, working in Bristol, have recently found extremely high levels of oxytocin in ovine corpora lutea (see Suggested Further Reading).

Some thoughts for the future

Our knowledge of the hypothalamic magnocellular neurones of the rat is arguably greater than that for any other neurone in the mammalian brain, given what we know of its structure, peptide synthesis, electrophysiology, secretion and pharmacology. Yet, as superficial as our knowledge is of other species, there is already sufficient evidence to indicate that many of the details presented, and even some of the principles elaborated, in this chapter may not apply throughout the mammalian class. In the future our research on the posterior pituitary has to be made more comparative (especially in the field of electrophysiology). Technically we do not face the same limitations that we did just 5 years ago. New anaesthetics are available that are more selective in the control of pain and stress, and using these we may soon be able to conduct experiments under anaesthesia that have hitherto been impossible. Techniques are now available that permit electrical recordings to be taken from single neurones in the conscious animal for periods of days, even in species as large as sheep. Indeed, the farm species lend themselves well to pharmacological studies involving the intraventricular administration of regulatory peptides and other agents. At the other extreme, it is now possible to culture *in vitro* the magnocellular neurones, potentially of any species, and so explore their biosynthetic and electrophysiological functions in exquisite detail.

On the other hand, our very considerable knowledge of the magnocellular neurones of the rat already permits ever more searching questions to be asked, and here there is still scope for everyone to participate. At the ultrastructural level, we need to know how the magnocellular nuclei are organized to permit the oxytocin and vasopressin cells to express their diverse activities, and we need to know how neurosecretory granules are constructed and how they are moved to and fro within the neurone. The biochemists among you have to resolve that all-important issue of what regulates the cleavage of the prohormone, and determine what regulatory peptides, if any, the magnocellular neurones produce in addition to oxytocin and vasopressin. The physiologist has the challenge of locating and deciphering the pulse generator that integrates the various sensory inputs for oxytocin release, which in turn sets the frequency for the pulsatile release of the hormone. Finally, the neuropharmacologist has for closer study the blood–brain barrier, which in the light of current evidence would appear to separate the central and peripheral actions of peptides; this suggests the exciting possibility that the key to many neuroendocrine processes may reside in the regulation of the peptidergic nerve terminal rather than in the synaptic control of the hypothalamic neurone itself.

Suggested further reading

Temporal patterns of neural activity and their relation to the secretion of posterior pituitary hormones. J. J. Dreifuss, E. Tribollet and M. Mühlethaler. *Biology of Reproduction*, **24**, 51–72 (1981).

Endocrine neurones. B. A. Cross, R. E. J. Dyball, R. G. Dyer, C. W. Jones, D. W. Lincoln, J. F. Morris and B. T. Pickering. *Recent Progress in Hormone Research*, **31**, 243–95 (1975).

The neurohypophysis. B. A. Cross and J. B. Wakerley. *International Reviews in Physiology*, **16**, 1–34 (1977).

Structure–function correlation in mammalian neurosecretion. J. F. Morris, J. J. Nordmann and R. E. J. Dyball. *International Reviews in Experimental Pathology*, **18**, 1–90 (1978).

The neurosecretory neurone: a model system for the study of secretion. B. T. Pickering. *Essays in Biochemistry*, **14**, 45–81 (1978).

Hormone receptor control of prostaglandin $F_{2\alpha}$ secretion by the ovine uterus. J. A. McCracken. *Advances in Prostaglandin and Thromboxane Research*, **8**, 1329–44 (1980).

Neuroendocrine control of milk ejection. D. W. Lincoln and A. C. Paisley. *Journal of Reproduction and Fertility*, **65**, 571–86 (1982).

Mammary gland growth and milk ejection in the agile wallaby, *Macropus agilis*, displaying concurrent asynchronous lactation. D. W. Lincoln and M. B. Renfree. *Journal of Reproduction and Fertility*, **63**, 193–203 (1981).

Electrophysiological evidence for the activation of supraoptic neurosecretory cells during the release of oxytocin. D. W. Lincoln and J. B. Wakerley. *Journal of Physiology*, London, **242**, 533–54 (1974).

Electrophysiological differentiation of oxytocin- and vasopressin-secreting neurones. D. A. Poulain, J. B. Wakerley and R. E. J. Dyball. *Proceedings of the Royal Society*, Series B, **196**, 367–84 (1977).

Is oxytocin an ovarian hormone? D. C. Wathes and R. W. Swann. *Nature*, **297**, 225–7 (1982).

Opiate inhibition of peptide release from the neurohumoral terminals of hypothalamic neurones. G. Clarke, P. Wood, L. Merrick and D. W. Lincoln. *Nature*, **282**, 746–8 (1979).

Calcium localization in the rat neurohypophysis. F. D. Shaw and J. F. Morris. *Nature*, **287**, 56–8 (1980).

Nucleotide sequence of cloned cDNA encoding bovine arginine vasopressin-neurophysin II precursor. M. Land, G. Schütz, H. Schmale and D. Richter. *Nature*, **295**, 299–303 (1982).

3

The pineal gland

GERALD A. LINCOLN

The pineal gland of mammals has had a checkered history. René Descartes is well known for his seventeenth century view that the pineal in man was the seat of the soul. He thought that the organ functioned like a tap in the centre of the brain, regulating the flow of vital fluids through the ventricles, and receiving information from the eyes and other sense organs (Fig. 3.1). Others have seen the pineal as of lesser importance, perhaps a mere functionless vestige left over from early vertebrate evolution. In man it is small and becomes progressively calcified, so it has perhaps been easy to assume that the gland is of no significance.

A French clinician Dr O. Heubner must be given the credit for linking the pineal gland with reproduction, and putting us on the right track. He presented observations in 1898, on a case of a 4-year-old boy with

Fig. 3.1. Seat of the soul. A drawing by René Descartes showing the seventeenth century view that the pineal gland plays a central role in the control of the brain, receiving information from the sense organs.

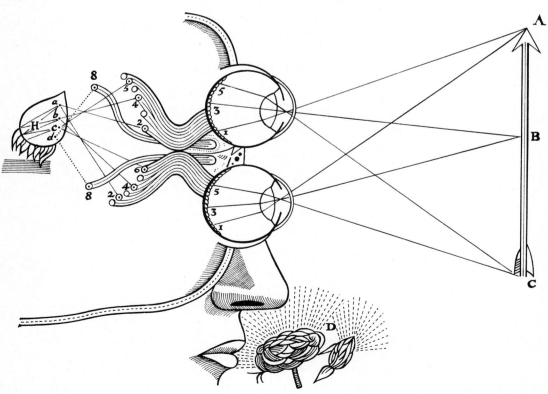

precocious puberty whom it was discovered had a pineal tumour. To explain the condition, Heubner speculated that the pineal must normally secrete a hormone that inhibits sexual development in children, and when its production is disrupted by disease puberty may then be precocious. This view was not very popular at the time and most clinicians preferred to associate the disturbance of pubertal development with the physical damage resulting from the pressure of an enlarging tumour. Logically this could influence the function of the hypothalamus and thus the control of the secretions from the anterior pituitary gland.

Then in 1954 came an important observation. Julian Kitay, working at the Harvard Medical School in Massachusetts, reviewed the incidence of pineal lesions and precocious puberty. He showed that there was a highly significant correlation between the histological appearance of a pineal tumour and the occurrence of abnormal sexual development; this could not be explained simply in terms of pressure effects elsewhere in the brain. The parenchymatous cells of the pineal appeared to be required for the normal timing of development at puberty. Within 5 years, Aaron Lerner had isolated the first pineal hormone which he called melatonin, and the era of pinealology had begun.

We now know that in mammals the pineal is a neuroendocrine gland and plays a significant role in regulating reproduction. It is important during puberty in the young animal, and in the switching on and off of reproduction according to season in the adult. The initial impression was that the pineal secretes inhibitory hormones that suppress reproduction, but evidence is growing that the pineal is geared into the circadian organization of the brain. It secretes its hormones in relation to night and day, and is thus involved in time keeping. Its role in regulating reproduction is concerned with timing events rather than causing them.

Evolution from sense organ to endocrine gland
The pineal has a long history. Fossils of jawless fish from the Silurian era, some 420 million years old, show a clearly defined though small pineal recess in the centre of the head (Fig. 3.2). The pineal of these earliest vertebrates apparently had some sort of visual contact with the outside world through the skin and so acted as a third eye. Its function was possibly related to the control of skin colour in camouflage, since these animals lived on the sea bed, in shallow water.

Fossil records provide few clues to the way the soft structures of the pineal became transformed during the evolution of the vertebrates. A possible sequence can be deduced, however, from the diverse anatomy of the pineals of some present-day fish, reptiles and amphibians. Many of these species have a complex pineal, differentiated in two distinct regions, the pineal organ and an anterior structure known as the parietal organ. The pineal organ appears to be the homologue of the mammalian pineal.

The evolution of the pineal cell of mammals seems to have involved a change from a light-sensitive cell to an endocrine cell; the inferred stages have been summarized by Jean-Pierre Collin working at Poitiers in France, and are shown diagrammatically in Fig. 3.3. The changes involved the loss of the terminal region of the cell specialized for photoreception and the loss of sensory nerve cells connecting it with other areas of the brain. Studies on the fetal development of the pineal gland of mammals show a transitory appearance of some of the primitive pineal structures including a pineal nerve, providing an example of the way ontogeny recapitulates

Fig. 3.2. How old is the pineal gland?. The jawless fish *Hemicyclaspis*, fossilized some 420 million years ago, had a pineal recess (P) on the upper surface of the head, indicating that the pineal organ had already reached an advanced stage of evolution in the earliest vertebrates. (Redrawn from E. A. Stensio. The Cephalaspids of Great Britain. *Brit. Mus. Nat. Hist.* 1–220 (1932).)

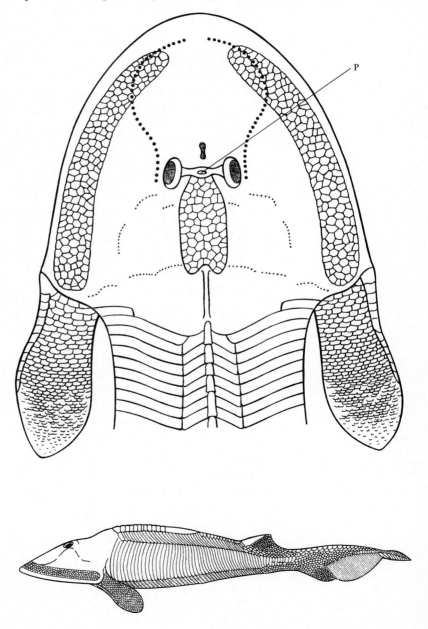

phylogeny. While the pineal of mammals is no longer a photosensory organ, it does change its function in relation to daylight and darkness, a feature that is dependent on innervation by sympathetic nerves. This feature is critical for its present-day endocrine function.

Anatomy and strange innervation

The pineal gland is the epiphysis cerebri of anatomists. The word 'pineal' comes from the Latin *pineus* meaning 'of the pine', and refers to the cone-shaped appearance of the gland in man. Because of its elongated form in some animals, it has also been called the penis of the brain. Amongst the different species of mammals there is much variation in size and shape of the pineal as can be seen by comparing man, rabbit, sheep and Weddell seal (Fig. 3.4); one of the biggest pineals is found in the elephant seal. There are a few exceptions, like the dugong and giant anteater, in which the pineal gland is apparently totally absent, and others like the elephant, rhinoceros and opossum in which it is small. Most mammals, however, have a clearly defined pineal which can be seen if the brain is removed from the skull

Fig. 3.3. Evolution of the pinealocyte. The pineal organ of some primitive vertebrates contained photosensory cells connected by nerves to the brain which constituted the third eye. During the course of mammalian evolution the pineal cells have become transformed into neurosecretory cells and the response to light has been lost. The sequence has been traced by a study of the pineals of present-day fish, reptiles and amphibians. b.v., blood vessel; g, Golgi complex; l.s., light sensory region; m, mitochondria; n, nucleus; s.c., sensory nerve cell; s.g., secretory granules. (Redrawn from J. P. Collin. *Colloques Internationaux CNRS* **266**, 393–407 (1976).)

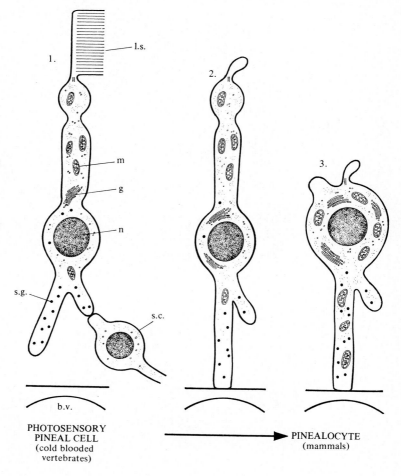

PHOTOSENSORY
PINEAL CELL
(cold blooded
vertebrates)

PINEALOCYTE
(mammals)

and the occipital lobes of the cerebral hemispheres pulled apart. The long, pendulous pineal of animals like the rat, hamster and rabbit attaches to the membranes (meninges) covering the brain. A general feature is that mammals from higher latitudes have larger pineals than those from the tropics. This appears to reflect a role of the pineal in the regulation of seasonal breeding; indeed, within animals from the temperate climates, like the European hare or the elephant seal, the pineal changes in size according to the time of year.

The parenchymatous cells forming the main tissue of the pineal are called pinealocytes. These have a cytoplasm with relatively little endoplasmic reticulum but a conspicuous Golgi apparatus and many mitochondria. Dense vesicles, assumed to contain secretory products, appear to originate from the Golgi region and leave the cell from specialized zones close to blood capillary vessels. The pineal has a rich blood supply and is outside the blood–brain barrier.

The nerve supply to the pineal gland is unusual for a structure that is part of the brain. Ariens Kappers was one of the first investigators to study the nerve supply in detail, working mainly with the rat. He concluded that in mammals there are no afferent or efferent nerve fibres directly linking the pineal with the other areas of the brain. Instead, the gland receives a peripheral innervation from sympathetic nerve fibres, which originate from the superior cervical ganglia in the neck region, and traverse along the arterial supply to the head before forming the nervi conarii that enter the pineal. Within the gland the fibres form a plexus with nerve terminals close to the capillary vessels and the pinealocytes. Some mammals such as the

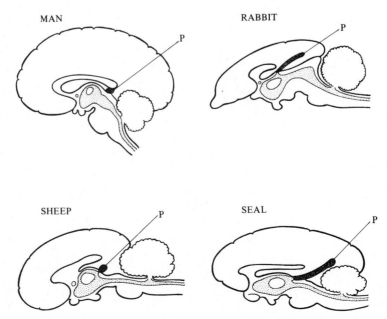

Fig. 3.4. Species variation. Amongst the mammals there is considerable variation in the size and shape of the pineal gland (P), as illustrated here for man, rabbit, sheep and Weddell seal. The largest pineals are found in species like the seal inhabiting cool, seasonal climates.

rhesus monkey and rabbit are said to have a parasympathetic nerve supply to the pineal in addition to that from the sympathetic system.

The link between the eyes and the pineal gland is through the sympathetic nerve supply. At least, this appears to be true from the functional point of view. For example, Robert Moore of the University of California has used the daily rhythm in the synthesis and secretion of the pineal indoleamines in the rat as an index of functional activity, and traced the nerve connections through the brain (Fig. 3.5). Photic information is transmitted from the retinal cells of the eye, via the optic nerves and connections in the base of the brain and brain stem, to the superior cervical ganglia, and thus to the pineal. The suprachiasmatic nucleus, lying above the optic chiasma in front of the hypothalamus, is a region of special importance in this relay since it appears to generate the daily rhythmic pattern of secretion by the pineal.

While the importance of the sympathetic innervation of the pineal must be emphasized, there is evidence of other links between the eyes and the pineal not involving the superior cervical ganglia. For example, changes in electrical activity can be picked up in the pineal gland of rats if the animals are given a stimulus of light directed at the eyes (Fig. 3.6). This

Fig. 3.5. Innervation of the pineal gland. The pineal gland in mammals is richly innervated by sympathetic nerve fibres arising from the superior cervical ganglia. Photic information from the eyes and suprachiasmatic nucleus arrives via this sympathetic innervation. (Redrawn from R. Y. Moore. *Prog. Reprod. Biol.* **4**, 1–29 (1978).)

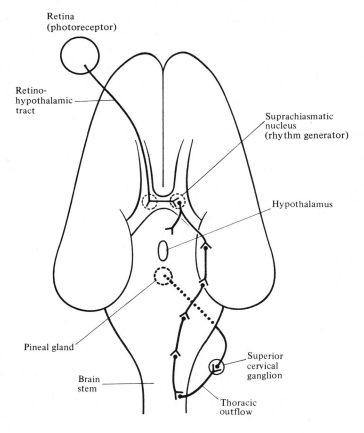

photic response in the pineal gland is only slightly modified when the rats have both superior cervical ganglia anaesthetized, thus preventing any neural transmission via this peripheral route. In spite of these observations of a direct link with the brain, the pineal appears to depend on its sympathetic innervation for its normal function.

Hormones of the pineal gland

In the last 25 years much effort has gone into attempts to isolate hormones from the pineal gland. There have been some notable successes, but many people believe that the most important compounds remain undiscovered. One of the main difficulties in this work has been the development of a suitable bioassay for use during the long isolation procedures. This is because there are uncertainties about the way the pineal gland normally functions. Does it affect reproduction by influencing the hypothalamus, or the anterior pituitary gland, or by acting directly on the testes or ovaries? Ironically, the initial breakthrough in the unravelling of the biochemistry of the pineal gland was the isolation of melatonin by Aaran Lerner and colleagues at Yale in 1959, using the rather bizarre frog-skin bioassay.

Isolation of melatonin

Lerner has described the isolation of melatonin as requiring an enormous effort. It depended on the close collaboration of several people, notably biochemists Yoshiyata Takahashi and James Case. Lerner was led to investigate the pineal through his interest in skin colour in people, with the hope that it might lead to treatment of conditions such as vitiligo, which leaves individuals with patchy pigmented skin, giving them a piebald appearance. He was aware of the work of McCord and Allen published

Fig. 3.6. Direct links with the brain. In this experiment with adult rats, a light stimulus (1) was applied to the eye while an electrical recording was made from electrodes (2) placed in the pineal gland. Anaesthesia of the superior cervical ganglia (3) only prevented part of the photic response, indicating that a direct link between the brain and pineal may exist. (From N. Dafny. *J. Neural Trans.* **48**, 203–11 (1980).)

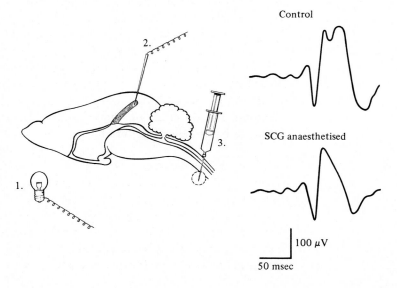

Control

SCG anaesthetised

100 μV

50 msec

in 1917 showing that extracts of the pineal gland can cause a lightening of colour of the skin of tadpoles. The question was – what was the active ingredient?

The starting material was many thousands of pineal glands collected from cattle at the local abattoir. The initial bioassay on the frog-skin was somewhat tricky, since it proved necessary to use melanocyte stimulating hormone to darken the frog-skin before using the preparation to detect the presence of the unknown lightening substance. Melanocyte stimulating hormone is a polypeptide secreted from the intermediate region of the pituitary gland, as a fragment of the large pro-opiocortin molecule (also the parent molecule for adrenocorticotrophic hormone and β-endorphin); it is not to be confused with the unrelated compound, melatonin.

Using a series of extractions and chromatographic separations, Lerner's team established that the skin-lightening compound from the bovine pineals was an indoleamine. With fluorescence and ultraviolet light absorption spectroscopy they were first able to isolate a substance very similar biochemically to the active ingredient but present in much larger quantities. This was purified and ultimately identified as 5-methoxyindole acetic acid.

At this point it became apparent that to get enough of the active skin-lightening compound for identification was going to be very time-consuming. Apparently over a million pineal glands would have been needed, and this at the time posed an impossible task. The decision was reluctantly made to drop the project. About a week before the final deadline however, Lerner was discussing what was known about the biosynthesis of indoleamines and an idea suddenly occurred to him. He guessed that the elusive compound in the pineal of cattle must be the methoxy derivative of *n*-acetylserotonin. Immediately his team set to work and within a day the first tests with the frog-skin assay were complete; never before had they tested anything so potent. The problem was solved. They could never have determined the structure of the lightening factor by direct attack: instead it took an inspired guess!

For a name, Lerner chose to call the active compound melatonin since it was related to serotonin and affected melanin pigmentation. As it has turned out the name is not particularly appropriate, since in mammals melatonin has no effect on the melanophores and its principal effects appear to be on cells of the brain.

Biosynthesis of melatonin
Since the isolation of melatonin, the biosynthesis of this indoleamine has been worked out in detail. The pineal gland is not the sole source of melatonin since it is synthesized in the retina, Harderian gland (in the eye orbit) and gut, at least in the rat. However, the pineal gland is the major producer. In addition, melatonin is not the sole indoleamine produced in the pineal gland, although it does seem to be the most biologically active.

The biosynthesis of melatonin has been investigated by many researchers including Richard Wurtman of the Massachusetts Institute of Technology and David Klein of the National Institutes of Health, again working mostly with rats. The precursor is the essential amino acid tryptophan which is taken up from the blood by the pinealocytes. The tryptophan is converted by hydroxylation and carboxylation to 5-hydroxytryptamine (serotonin) in two steps catalysed by specific enzymes (Fig. 3.7). The concentration of serotonin in the pineal is often higher than in any other tissue in the body.

Serotonin is converted to melatonin in two further stages, the first involving the enzyme *N*-acetyl transferase which in the rat provides an important rate-limiting step. The final enzyme is hydroxyindole-*o*-methyltransferase which converts *N*-acetyl serotonin to melatonin. Melatonin is not stored in large amounts in the pinealocytes, so its synthesis closely reflects the pattern of secretion and its concentration in the peripheral blood. Once released, it is rapidly metabolized by the liver, with a short half-life in blood of less than 10 minutes. Some melatonin enters the cerebrospinal fluid, where its fate is likely to be different, and survival longer.

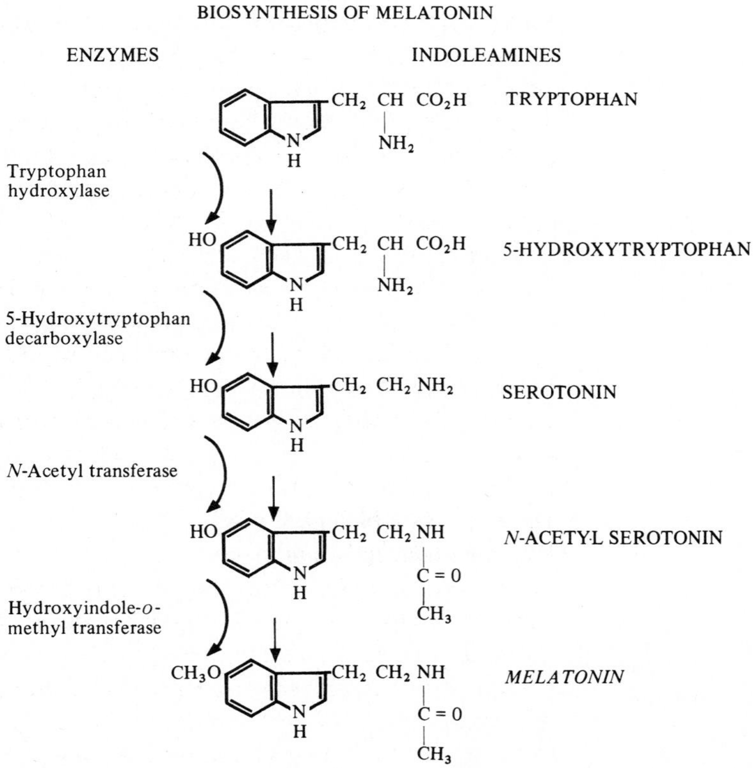

Fig. 3.7. Biosynthesis of melatonin. Melatonin is synthesized from the essential amino acid tryptophan in four steps each catalysed by a specific enzyme. The pineal gland is the primary source of melatonin, although some is produced in the retina, Harderian gland and gut. (From R. J. Wurtman, and M. A. Moskowitz. *New England J. Med.* **296**, 1329–33 (1977).)

BIOSYNTHESIS OF MELATONIN

ENZYMES   INDOLEAMINES

TRYPTOPHAN

Tryptophan hydroxylase

5-HYDROXYTRYPTOPHAN

5-Hydroxytryptophan decarboxylase

SEROTONIN

N-Acetyl transferase

N-ACETYL SEROTONIN

Hydroxyindole-*o*-methyl transferase

MELATONIN

Peptide hormones versus indoleamines

The success in unravelling the biosynthesis of the pineal indoleamines has rather overshadowed the attempts to isolate the peptide hormones produced by the gland. However, much research effort has gone into studying these compounds since there is a well-founded belief that the peptides represent the true hormones that influence reproduction. The peptides identified or partially purified so far include arginine vasotocin, pineal anti-gonadotrophin, gonadotrophin releasing hormone (pineal GnRH and different from hypothalamic GnRH), prolactin releasing and inhibitory factors, and thyrotrophin releasing and inhibitory factors. The wide spectrum of compounds may reflect the many functions of the pineal gland, or simply the enthusiasm of the scientists involved in the search!

The pineal peptide that has received most attention is arginine vasotocin. This was first isolated from pineal glands in 1963 by Stephan Milcu and his colleagues, and its identity was proven chemically in 1970 by Dean Cheesman and Bruce Fariss. Much interest has surrounded this putative hormone since it is much more effective in suppressing the secretion of gonadotrophic hormones in mammals than is melatonin. The extreme potency of arginine vasotocin in this respect is highlighted in a recent publication in which it was shown that a minute dose of 100 attograms (10^{-16} g) injected into the third ventricle of the brain of a castrated rat led to depression in the secretion of LH. Supporters of the arginine vasotocin hypothesis claim that the dense vesicles in the pinealocytes contain this compound which is released into the blood or cerebrospinal fluid. In this respect we could say that the pineal functions somewhat like the posterior pituitary gland which secretes the closely related peptides arginine vaso-pressin and oxytocin. A neurophysin-like protein has been identified in the pineal gland which may represent part of a prohormone, as in the posterior pituitary gland (see Chapter 2).

Work on the other pineal peptides, including pineal antigonadotrophin, has produced few conclusive results. This is because their amino acid composition has not been fully defined and the peptides are not yet available in purified form for experimental use. One tripeptide, threonine–serine–lysine has been isolated recently by Richard Orts and his colleagues, but this compound has not proved to be biologically active when tested by others.

The pineal as a biological clock

Day and night, light and dark

There is a very pronounced daily rhythm in the secretory activity of the pineal gland. This is well documented for man, as well as for various experimental animals such as the rat, rhesus monkey and sheep, from measurements of the levels of melatonin circulating in the blood. The levels of melatonin are highest at night in all the species irrespective of whether they are nocturnal or diurnal in their activity patterns. An example of the melatonin rhythm is illustrated in Fig. 3.8 for Soay sheep which were kept under an artificial lighting cycle. A similar pattern is found in man living under natural conditions, with peak blood levels usually occurring at night. In the rat there is a daily rhythm in the blood levels of arginine vasotocin similar to that for melatonin, indicating that the secretion of the peptides from the pineal may change in close parallel with that of the indoleamines.

Fig. 3.8. 24 hour rhythms in melatonin. Changes in the concentration of melatonin measured by radioimmunoassay in the peripheral blood of eight Soay rams exposed initially to a 24 h light/dark cycle (long day of 16 h light:8 h darkness) and then to 24 h of darkness. Note how the daily rhythm in the levels of melatonin persists during prolonged darkness. (From O. F. X. Almeida and G. A. Lincoln, unpublished results, 1981).

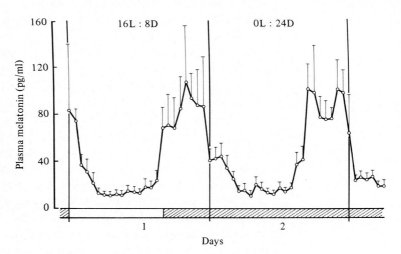

Fig. 3.9. Night-time surge in melatonin. Studies on the rat indicate that the increase in the secretion of melatonin which normally occurs at night results from stimulation of the sympathetic nerve fibres in the pineal gland (1). This results in release of noradrenaline (2) which stimulates the β-adrenergic receptors on the cell membrane of the pinealocytes, releasing cyclic AMP which acts as second messenger within the cell (3). The next step is an increase in the synthesis of n-acetyl transferase (4) which increases 10-fold in activity at night, and appears to be the rate-limiting step in the synthesis of melatonin (5). Serotonin normally accumulates during day-time, and is converted to melatonin at night by N-acetyl transferase and hydroxyindole-o-methyl-transferase.

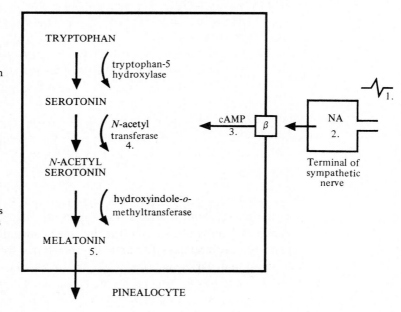

In sheep it has been shown that there is a distinct change in the blood flow through the pineal with a rhythm parallel to the rhythm in the secretion of melatonin.

The daily rhythm in the activity of the pineal gland is dictated by the activity of the sympathetic nerves supplying the gland. Recordings from the pineal cells of rats have shown an increase in electrical activity at the onset of darkness. This results in the release of noradrenaline at the nerve terminals close to the pinealocytes, and on the neighbouring capillaries. This neurotransmitter stimulates the biosynthesis of melatonin, and possibly the peptide hormones, through binding to the β-adrenergic receptors located in the cell membrane of the pinealocytes (Fig. 3.9).

The biochemical events by which noradrenaline has its effects on the melatonin system have been studied by using tissue cultures of pineal glands as originally described by Harvey Shein and his colleagues in 1967. When rat pineals are cultured in a medium containing isotopically labelled tryptophan, they convert the amino acid to labelled melatonin and release this into the medium at a fairly linear rate for at least 48 h. The addition of noradrenaline to the cultures stimulates melatonin synthesis while serotonin and melatonin itself are without effect. The stimulatory effect of noradrenaline on the cultured cells is prevented by drugs such as propanolol which act as β-receptor blockers. Also, the stimulatory effect is mimicked by dibutyryl cyclic AMP which is assumed to be the second messenger conveying the stimulus from the receptor on the cell membrane to the nucleus to initiate the response.

The effect on the cell is to stimulate the synthesis and activation of enzymes involved in the biosynthesis of melatonin. The regulation of the enzyme *N*-acetyl transferase, which catalyses the conversion of serotonin to *N*-acetyl serotonin, is claimed to be particularly important as a rate-limiting step; this appears to be the case since the enzyme shows a 20-fold increase in activity in response to stimulation by noradrenaline, while the activity of other enzymes in the pathway such as hydroxyindole-*o*-methyltransferase changes only slightly (Fig. 3.9).

Circadian rhythms and dependence on the suprachiasmatic nucleus
The impression given so far is that the daily rhythm in the secretory activity of the pineal gland is directly dictated by the daily cycle of daylight and darkness. This is not strictly true, however, since the daily pineal rhythm persists for several days if an animal is transferred from a normal light cycle to continual darkness. This is illustrated in Fig. 3.8 which shows the melatonin levels in Soay sheep exposed to darkness after a long daily photoperiod. In darkness, the pineal melatonin rhythm free-runs with a period of about 24 h, i.e. a *circadian rhythm*.

There are two ways in which the rhythm in pineal activity could be generated endogenously. The gland could have an intrinsic rhythm in secretory activity or it could be driven to show a rhythm in activity from elsewhere in the body. In mammals, the second alternative appears to be correct since the isolated mammalian pineal gland in organ culture does not have inherent rhythmic properties. This is unlike the situation in some species of birds, like the chicken. The regions of the brain that apparently dictate the activity of the pineal gland in mammals are the suprachiasmatic nuclei located bilaterally in the anterior hypothalamus immediately above the optic chiasma. In the rat at least, these small areas consisting of a few thousand neurones appear to be the circadian generator for the entire body, controlling such things as the daily sleep/wake cycle, the daily rhythm in body corticosterone, temperature and heart rate, and the pineal rhythm in melatonin secretion. Destruction of these nuclei produces an animal which is arhythmic, while operations that disconnect the nuclei from the pineal gland leave most systems intact but block the pineal's circadian melatonin rhythm (Fig. 3.10).

While the endogenous pineal rhythm is controlled by the suprachiasmatic nuclei, it is normally entrained to a particular time of day by the prevailing daily light cycle. In this respect the eyes act as the photoreceptor and the information is conveyed to the nuclei along the optic nerves as mentioned earlier. The pathway is separate from the visual system and is known as the retino-hypothalamic tract. The circadian rhythm in the activity of the pineal persists if the eyes are removed or if the optic nerves are cut, just as when animals are kept under constant darkness (Fig. 3.10). The

Fig. 3.10. Regulation of 24-hour rhythms in the pineal. This summarizes the experimental procedures that disrupt the 24 h rhythm in indoleamine biosynthesis in the pineal gland. The rhythm persists following removal of the eyes, cutting of the optic tract, or in normal animals under constant darkness. It is disrupted by any procedure interfering with the neural link between the suprachiasmatic nucleus (SCN) and the pineal gland, as well as under constant light. SCG, superior cervical ganglion. (From R. Y. Moore. *Prog. Reprod. Biol.* **4**, 1–29 (1978).)

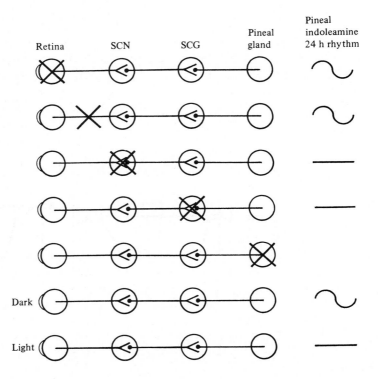

circadian control of the pineal gland determines how the gland influences reproduction, and we will return to this in the next section.

Roles in reproduction

Seasonal breeding – switching on and switching off

The role of the pineal gland in the control of reproduction has been most thoroughly studied in seasonally breeding mammals. The golden hamster was the first species to be investigated. This small rodent from Syria has a pronounced annual reproductive cycle; the testes of the males are largest in spring, and at this time the females come into oestrus. It is a photoperiodic species, in which the seasonal changes in day-length govern the timing of the seasonal reproductive cycle.

In 1964 a group of French scientists discovered that removal of the pineal gland in hamsters converts the animal from a seasonal to a continuous breeder. This effect has since been studied in great detail by Russell Reiter, of the University of Texas, and many others. A simple experiment is summarized in Fig. 3.11. When adult male hamsters are transferred from long days to short days (> 12.5 h light per day is a long day for hamsters) their testes regress to about half their normal size in 10 weeks. If the pineal gland is removed and the animals are then exposed to the change in photoperiod, the regression of the testes does not occur. Similarly, pinealectomy abolishes the inhibitory effects of constant darkness or blinding on the reproductive system of the hamster. Denervating the pineal gland by superior cervical ganglionectomy, or destroying the suprachiasmatic nuclei, has the same effect as pinealectomy.

Fig. 3.11. Effects of pinealectomy on reproduction. Removal of the pineal gland, or its denervation by removal of the superior cervical ganglia, disrupts the normal photoperiodic control of reproduction in seasonally breeding animals. For example, the male hamster does not show regression of the testes in response to a change from long to short days after pinealectomy (*a*), while the Soay ram is equally insensitive to photoperiodic manipulations after superior cervical ganglionectomy (*b*). PINX, pinealectomy; SCGX, removal of superior cervical ganglion. (From F. Turek and C. S. Campbell. *Biol. Reprod.* **20**, 32–50 (1979); G. A. Lincoln. *J. Endocr.* **82**, 135–47 (1979).)

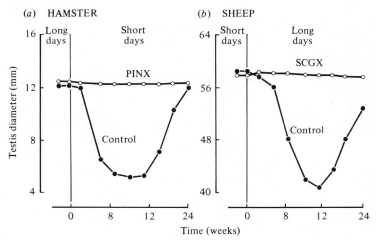

(a) HAMSTER   *(b)* SHEEP

Time (weeks)

Studies on the Soay sheep carried out by my own group in Edinburgh have shown that denervation of the pineal gland affects reproduction in this species also. Sheep differ from hamsters in being switched off by long days instead of short days, but in both species an intact pineal gland is required for the normal photoperiodic response (Fig. 3.11).

From these observations we infer that functional pinealectomy converts a seasonal breeder into a continuous breeder. This is not strictly true, however, in the case of longer-lived species like the sheep, deer and ferret. For example, pinealectomized sheep and deer kept under natural lighting and temperature continue to have annual breeding seasons, although these do not necessarily occur at precisely the normal time. Since it is known from experiments with controlled lighting that removing the pineal renders sheep non-photoperiodic, other seasonal cues from the environment would appear to influence the timing of the breeding cycle. Proof that seasonal periodicity can persist after pinealectomy even with no cues from the environment is well illustrated by the work of Joe Herbert and his colleagues on ferrets at the University of Cambridge (Fig. 3.12). When pinealectomized ferrets are housed under natural lighting but constant temperature they continue to have periodic breeding seasons for up to four years. These breeding periods occur at irregular intervals with no synchrony around any particular time of year. The effect of pinealectomy is similar

Fig. 3.12. Timing the breeding season. In the ferret the pineal gland is required if the breeding season is to occur at the appropriate time of year. The diagram shows the periods of oestrus in control and pinealectomized ferrets living under natural lighting. X = Animal died. (From J. Herbert, P. M. Stacey and D. H. Thorpe. *J. Endocr.* **78**, 389–97 (1978).)

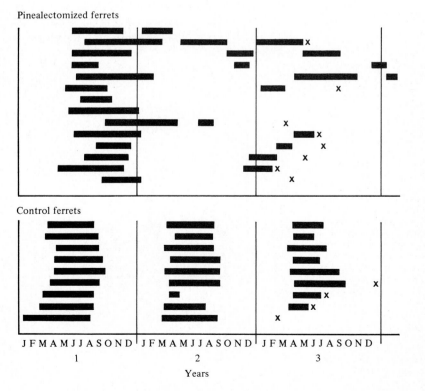

Pinealectomized ferrets

Control ferrets

J F M A M J J A S O N D   J F M A M J J A S O N D   J F M A M J J A S O N D
          1   2   3

Years

to that of blinding since the ferrets ignore the seasonal changes in photoperiod.

The lesson to be learned from these observations is that the pineal gland is not simply involved in the inhibition of reproduction to produce anoestrus and infertility. The pineal gland is involved in the timing of the seasonal cycle in reproduction. The gland is involved in relaying the effects of photoperiod on reproduction, and since changes in photoperiod can stimulate as well as inhibit reproductive activity, it follows that the pineal can appear to be either stimulatory or inhibitory. The significance of this will become more apparent when we consider the effects of melatonin on reproduction.

The measurement of melatonin in the blood of animals under different photoperiods provides a way of seeing how the pineal gland might influence reproduction. One example is given in Fig. 3.13 for Soay rams housed under either short days, when they become sexually active, or under long days when they became sexually quiescent. Under both lighting conditions the blood levels of melatonin are increased at night: however the night-time peak in melatonin changes in amplitude and duration, and occurs at a different time relative to 'dawn' under the two photoperiods. These differences in the pattern of melatonin secretion may thus relay photoperiodic effects and cause the changes in testicular activity.

Puberty – the first breeding season

The first evidence for the involvement of the pineal gland in the control of reproduction came from observations on puberty in boys. As mentioned

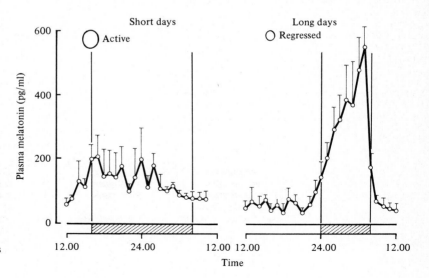

Fig. 3.13. Does photoperiod influence reproduction through effects on the pattern of melatonin secretion? This shows the changes in the blood levels of melatonin in four adult Soay rams sampled over 24 hours during exposure to 16 weeks of short days when sexually active, and over 24 hours during exposure to 16 weeks of long days when sexually inactive. The large increase in the levels of melatonin at night during long days may be important in causing the inhibition of reproduction. Ovals at top represent testis sizes. (Data from G. A. Lincoln, O. F. X. Almeida, H. Klandorf and R. Cunningham. *J. Endocr.* **92**, 237–50 (1982).)

at the beginning of this chapter, the abnormal function of the pineal gland in young children can result in puberty occurring at the wrong time. Recently it has become possible to measure the blood levels of melatonin in children to see how the pineal behaves during puberty. In one such study the blood levels of melatonin were found to be high in young boys and to decline as puberty advanced. It has also been suggested that the darkening of the hair in children is related to the decline in the secretion of melatonin.

Experiments with animals have confirmed the involvement of the pineal gland in the control of puberty. For example, when female rats are pinealectomized within 5 days of birth, they show premature enlargement of the ovaries and uterus with earlier vaginal opening (the conventional index of puberty in the rat). In male rats, removal of the pineal at an early age can lead to precocious growth of the seminal vesicles and prostate, and early development of spermatogenesis in the testes. These effects of pinealectomy can be reversed if the experimental animals are treated with extracts of pineal glands.

Not all manipulations of pineal activity in young animals result in precocious puberty. In the white-tailed deer, for example, pinealectomy of fawns results in a delay in the development of the antlers and other secondary sexual characteristics. In the Soay sheep, superior cervical ganglionectomy of 2-month-old lambs results in an initial acceleration of testicular development, followed by premature regression before the normal breeding season (Fig. 3.14). In these last two examples, the timing of puberty is influenced by the prevailing photoperiod, and thus the pineal may only be influencing the timing of the photoperiodic response as in seasonal breeders, rather than having a direct effect on puberty itself. However, these two effects probably involve a common pathway of control and may be similar.

Fig. 3.14. Puberty in a seasonal breeder. The diagram shows the effect of removing the superior cervical ganglia (SCGX) from Soay ram lambs at the age of 2 months on the subsequent growth and regression of the testes. The SCGX lambs showed accelerated testicular enlargement and regression. (G. A. Lincoln, unpublished data.)

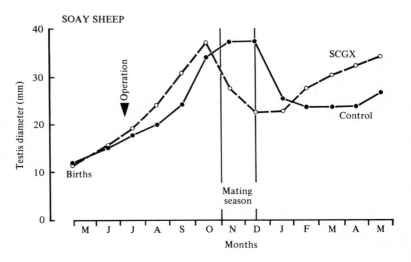

Mechanisms of action

In many of the studies on the role of the pineal gland in seasonal breeding and puberty, the change in the blood levels of LH, FSH and prolactin have been measured. The usual finding is that there is a close correlation between any change in the function of the reproductive organs and the gonadotrophin titres. Thus, it follows that the pineal gland must influence reproduction by affecting the secretion of hormones from the anterior pituitary gland. This may be a direct effect on the pituitary, or an indirect effect resulting from changes elsewhere in the brain. For example, effects on the hypothalamus could influence the secretion of releasing and inhibitory hormones. The mechanism of control by the pineal gland is only just beginning to be understood.

Experiments with melatonin

Many experiments have been performed to establish whether melatonin is the pineal hormone that has an influence on reproduction. The most useful animal model has been the golden hamster, since the role of the pineal gland in the photoperiodic response of this species has been well established. One experiment involving the daily injection of 25 μg melatonin into female hamsters is summarized in Fig. 3.15. In this case the animals were kept under long days and were showing regular oestrous cycles typical of the breeding season. Melatonin treatment caused a suppression of oestrous cyclicity beginning within 4 weeks and rendering all animals anoestrous within 7 weeks. This effect also occurs in male hamsters, where regression of the testes occurs; this has been shown to result from an

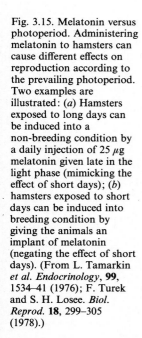

Fig. 3.15. Melatonin versus photoperiod. Administering melatonin to hamsters can cause different effects on reproduction according to the prevailing photoperiod. Two examples are illustrated: (*a*) Hamsters exposed to long days can be induced into a non-breeding condition by a daily injection of 25 μg melatonin given late in the light phase (mimicking the effect of short days); (*b*) hamsters exposed to short days can be induced into a breeding condition by giving the animals an implant of melatonin (negating the effect of short days). (From L. Tamarkin *et al. Endocrinology,* **99**, 1534–41 (1976); F. Turek and S. H. Losee. *Biol. Reprod.* **18**, 299–305 (1978).)

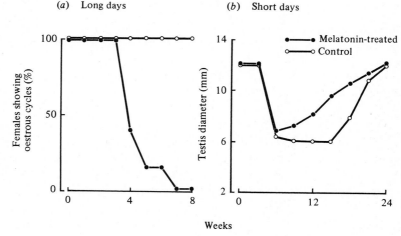

(*a*)   Long days   (*b*)   Short days

Females showing oestrous cycles (%)

Testis diameter (mm)

●——● Melatonin-treated
○——○ Control

Weeks

inhibition of gonadotrophin secretion. An additional observation is that the effect of melatonin depends on the *time of day* at which the injection is given. Thus, 25 μg melatonin given in the animal's subjective afternoon inhibits reproductive activity, while 25 μg melatonin given in the subjective morning is without effect. This effect is not evident in pinealectomized hamsters, indicating that the exogenous melatonin summates with the endogenous hormone to produce this effect.

These results can be compared to what happens when hamsters are transferred from long days to short days; the effect is similar to that produced by treatment with melatonin in the afternoon. We have already seen that the response to short days in the hamster is dependent on the pineal gland, and that melatonin secretion increases in darkness. The obvious conclusion therefore is that under short days the longer period of high melatonin levels associated with the long nights in some way causes an inhibition of gonadotrophin secretion and hence infertility.

The picture can appear to be even more complicated. In some situations in the hamster, melatonin appears to *stimulate* reproduction rather than inhibit it. One such experiment is illustrated in Fig. 3.15. In this case male hamsters were exposed to a prolonged period of short days; this treatment resulted in regression of the testes followed by spontaneous redevelopment after some 18 weeks as the animals became refractory to the inhibitory photoperiod. When some of the hamsters were given implants of melatonin at the time of full testicular regression, this resulted in premature redevelopment of the testes due to an increase in gonadotrophin secretion (Fig. 3.15). This has been called the pro-gonadal effect of melatonin, and when first described caused much confusion. However, the simple interpretation is that the exogenous melatonin merely nullifies the effect of the endogenous melatonin and hence the influence of the prevailing photoperiod. In this example, the animals were exposed to short days which normally suppress reproduction. We have already concluded that it is the pattern of melatonin secretion that is involved in the short day response, and so anything that modifies this pattern will disrupt the photoperiodic control of the seasonal cycle.

Besides the work on hamsters, the effects of melatonin on reproduction have been recorded in many mammals including ferrets, weasels, hares, rabbits, monkeys, rats, gerbils, goats and ponies. In one study Fred Turek and colleagues, working at Northwestern University, Illinois, found that melatonin treatment induced testicular atrophy in two photoperiodic species, the golden hamster and grasshopper mouse, but had no effects on gonadal size in two non-photoperiodic species, the laboratory rat and house mouse. This supports the idea that melatonin is a photoperiodic hormone – it is secreted by the pineal gland in relation to the daily light–dark cycle in all species but only influences hypothalamic function in species that show seasonal breeding. These photoperiodic species must have a brain that is responsive to changes in the pattern of melatonin secretion.

Effects of melatonin on the brain

The regulatory effects of melatonin on reproduction result from changes in the secretion of LH, FSH and prolactin by the anterior pituitary gland. The control exerted by melatonin is probably at the level of the hypothalamus and in many ways the hormone can be considered as acting like an anaesthetic.

Specific, high affinity receptors for melatonin have been located in the brains of cattle, with highest concentrations in the medial basal hypothalamus. This region contains the neurones that secrete the GnRH controlling the secretion of LH and FSH by the anterior pituitary gland. Electrophysiological studies carried out on the rat have shown that melatonin influences the electrical activity of the GnRH neurones.

More widespread effects of melatonin in the brain also occur that involve changes in neuronal activity, and synthesis of neurotransmitters. Changes in serotonin and γ-aminobutyric acid in the cortex and various other regions of the brain of rats have been measured after treatment with melatonin. In man, melatonin produces changes in the electroencephalogram and induces sleep. We do not know how these effects might influence reproduction, but it is interesting that in children during puberty luteinizing hormone secretion is increased during sleep.

We have already mentioned that the effectiveness of exogenous melatonin varies according to the time of day it is administered. This rhythm in responsiveness is apparently dictated by the suprachiasmatic nucleus, since lesions in this region of the brain of the hamster render the animals unresponsive to treatment with melatonin; furthermore, microinjections of melatonin placed in this area of the hypothalamus in the white-footed mouse are effective in switching off reproduction if the injections are given at a specific time of day. This means that the nucleus is a key region in the control of reproduction by the pineal gland, for it governs not only the rhythm of responsiveness *to* melatonin, but also the daily rhythm in secretion *of* melatonin by the pineal gland. The suprachiasmatic nucleus is therefore the central biological clock that dictates the photoperiodic response via its slave, the subservient pineal gland, which produces one of the regulator hormones.

Model integrating melatonin and pineal peptides

A model in which melatonin is the principal pineal hormone is depicted in Fig. 3.16. An alternative model is also shown in the same diagram – the peptide model – in which melatonin merely plays a local role within the pineal gland, while the functionally active hormones are peptides secreted into the blood and cerebrospinal fluid. When isotopically labelled melatonin is injected into the blood of animals it is taken up in large amounts by the pineal gland. There are several publications showing that the action of melatonin on reproduction depends on the presence of the pineal gland. In organ cultures melatonin increases the incorporation of amino acids into proteins in the pinealocytes, and leads to an increase in the amount of colchicine bound to microtubular proteins involved in cell secretion. It also stimulates ultrastructural changes in the pineal cells similar to those occurring in animals kept in darkness; for example, there is a decrease in the number of secretory granules in the polar region of the pinealocytes. Since these granules are thought to contain the peptide hormones, melatonin could be acting as a releasing hormone for secretory products. There is also *in vivo* evidence to show that a release of the granules from the pineal

Fig. 3.16. Regulation by the pineal. Two models have been proposed to explain the influences of the pineal gland on reproduction. *Melatonin model*: The pineal gland secretes melatonin in relation to the daily light–dark cycle and this acts on specific areas of the hypothalamus (1 and 2) to influence the secretion of the hypothalamic peptides that regulate the secretion of hormones from the anterior pituitary gland. *Peptide model*: The pineal gland secretes peptides including the pineal anti-gonadotrophin (PAG) and arginine vasotocin (AVT) which act on the hypothalamus and the pituitary (1 and 2) to influence the secretion of hormones from the anterior pituitary gland. The indoleamines of the pineal gland act locally in the control of the release of these peptides. MEL, melatonin; PRL, prolactin; CSF, cerebrospinal fluid; SCN, suprachiasmatic nucleus.

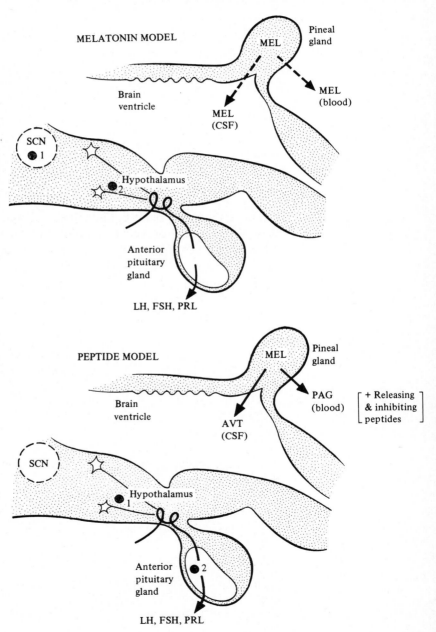

cells occurs at the same time as the increase in the synthesis and release of melatonin.

The weakness of the peptide model is that the peptide hormones have not been fully characterized. Arginine vasotocin may be involved, although many believe that the important hormone is still to be discovered. This is one of the most exciting areas for future research.

Practical aspects and conclusions

Animal breeding

In Edinburgh we first coined the term 'a ram for all seasons' to describe a ram that was functionally pinealectomized and unable to switch off its reproductive activity in relation to photoperiod. Since reproduction in the ewe is also under the control of day-length, we thought there may be a way of making sheep breed throughout the year. Unfortunately, this has not produced any practical developments. One of the difficulties is that sheep living outside continue to have periods of anoestrus even after pinealectomy, although they are not so pronounced as in the intact animal. Also the operation interferes with other seasonal changes, such as the cycle in appetite and the cycle in growth and moulting of the wool. There are various other seasonal breeding animals of commercial importance including the horse, red deer and mink, and control of pineal function in these species could prove to be useful. For example, in the mink it has recently been shown that melatonin implants can be used to synchronize the development of the valuable winter coat. The pineal gland is also involved in body growth and sexual development at puberty, and these could be modified. Immunization against the pineal hormones is one possibility for the future.

Human medicine

Abnormal functioning of the pineal gland in man may be clinically important. Accelerated or delayed puberty can result from a tumour of the pineal gland. More subtle malfunctions are probably more common and go undiagnosed at present. For example, some forms of secondary amenorrhoea in women may result from changes in the activity of the pineal gland; these changes could be similar to those occurring naturally in seasonal breeding animals which render them infertile for the non-breeding season.

A French clinician, Dr H. Laborit, has published a short account of some trials he carried out over 30 years ago which may illustrate the clinical importance of the pineal gland. He was a consultant at a hospital in Tunisia (Bezerte) and saw many cases of delayed puberty and amenorrhoea. His approach was to give a local injection of procaine into the superior cervical ganglion on both sides of the neck, thereby temporarily blocking the sympathetic innervation of the head. This was repeated two or three times per week for several weeks and in some cases it had distinct effects on sexual development. In one boy with delayed puberty, the treatment resulted in the enlargment of the testes, growth of pubic hair and appearance of spontaneous penile erections. At the time, Laborit believed the effects were due to changes in blood flow in the brain following changes in the sympathetic nervous control of the blood vessels. However, he later realized that the effects were probably due to alterations in the activity of the pineal gland and he published the results accordingly.

Today, measurement of melatonin and related indoleamines in blood and urine is beginning to be used as a diagnostic tool. Also, a pineal function test has been developed involving the injection of laevo-dopa, which stimulates the synthesis and release of melatonin. Another possible outcome of pineal research is the development of a new contraceptive. At present several research groups are intensively involved in the search for new peptide hormones in the pineal gland, which may specifically influence reproduction. There is reason to be optimistic since seasonal breeding is one of nature's contraceptives and its control involves the pineal gland.

The pineal gland plays an important role in the control of reproduction in mammals. Its effects are seen particularly during puberty and in seasonal breeding, and involve its action as a neuroendocrine transducer that converts neural information from the eyes into an endocrine message aimed at the hypothalamus. The pineal is intimately involved in time perception; it acts as a cog in the clock that measures the changes in the day-length, and thereby senses the seasons and regulates reproduction. Our knowledge is far from complete, however, and in the years to come we hope to define more precisely *which* of the pineal's products is the true biologically active hormone and *how* it acts within the brain to regulate ovarian and testicular activity.

Suggested further reading

Current status of pineal peptides. B. Benson. *Neuroendocrinology*, **24**, 241–58 (1977).

The sympathetic superior cervical ganglia as peripheral neuroendocrine centers. D. P. Cardinali, M. I. Vacas and P. V. Gejman. *Journal of Neural Transmission*, **52**, 1–21 (1978).

The photoneuroendocrine control of seasonal breeding in the ewe. S. J. Legan and S. S. Winans. *General and Comparative Endocrinology*, **45**, 317–28 (1981).

Seasonal breeding; nature's contraceptive. G. A. Lincoln and R. V. Short. *Recent Progress in Hormone Research*, **36**, 1–52 (1980).

Circadian rhythms of melatonin release from individual superfused chicken pineal glands *in vitro*. J. S. Takahashi, H. Hamm and M. Menaker. *Proceedings of the National Academy of Sciences of the United States of America*, **77**, 2319–22 (1980).

Photoperiodic regulation of neuroendocrine–gonadal activity. F. W. Turek and C. S. Campbell. *Biology of Reproduction*, **20**, 32–50 (1979).

The pineal organ. R. J. Wurtman and M. A. Moskowitz. *New England Journal of Medicine*, **296**, 1329–33 and 1383–6 (1977).

Physiological problems of seasonal breeding in eutherian mammals.
  J. R. Clarke. In *Oxford Reviews of Reproductive Biology*, pp. 244–312. Ed. C. A. Finn. Clarendon Press; Oxford (1981).

Biological Clocks and Seasonal Reproductive Cycles Colston Papers No. 32. Ed. B. K. Follett and D. E. Follett. John Wright; Bristol (1981).

The pineal and its hormones in the control of reproduction in mammals.
  R. J. Reiter. *Endocrine Reviews*, **1**, 109–31 (1980).

The pineal gland: a regulator of regulators. R. J. Reiter. *Progress in Psychobiology and Physiological Psychology*, **9**, 323–55 (1980).

Neural regulation of circadian rhythms. B. Rusak and I. Zucker. *Physiological Reviews*, **59**, 449–526 (1979).

Bibliography on photoperiodism in vertebrates, no. 123. B. K. Follett and F. W. Turek. *Bibliography of Reproduction*, **38** (1981).

4

The testis

D. M. de KRETSER

In the earlier chapters in this book, the complex functions of the hypothalamus and anterior pituitary have been considered. The concept that the testis is dependent on the function of the pituitary gland arose from the experiments of Philip Smith between 1920 and 1930. He was able to show that the size of the testis declined rapidly after removal of the pituitary gland (Fig. 4.1), and that this procedure affected the function of both compartments of the testis, the seminiferous tubules producing spermatozoa and the intertubular tissue containing the Leydig cells which produce steroid hormones, principally testosterone. Soon after this, Roy Greep and his colleagues demonstrated the existence of two anterior pituitary hormones, follicle stimulating hormone (FSH) and luteinizing hormone (LH), which when administered to hypophysectomized rats could reverse the atrophy of the testis. FSH acted on the seminiferous tubules and LH on the Leydig cells. Subsequent work with purer hormones has confirmed these early concepts, but more recent investigations suggest that the two compartments are not functionally independent and that there is in fact a close and complex interrelationship between them.

From the concepts developed in Chapter 1 it is clear that the hypothalamus plays an important role in controlling testicular function by secreting variable amounts of the gonadotrophin releasing hormone

Fig. 4.1. Decline in testicular weight of rats following hypophysectomy at different ages. (From R. Ortavant and M. Courot. *La Physiologie de la Reproduction chez les Mammifères*. Masson et Cie; Paris. (1980).)

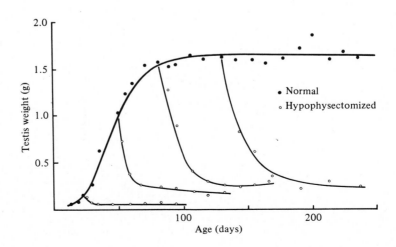

(GnRH). The hypothalamus serves as the link enabling the central nervous system to influence reproduction. By this mechanism environmental cues may exert their influence on reproductive processes and the seasonal control of reproduction in many mammals affords an excellent example of the mechanisms by which the hypothalamus and pituitary modulate testicular function. The detailed studies of Gerald Lincoln have shown that in the ram, day-length influences the hypothalamic secretory pattern of GnRH, which in turn alters the pattern of secretion of FSH and LH. During long days, basal FSH and LH levels are low, and pulsatile secretion of LH is very infrequent. However, with decreasing day-length, the increasing frequency and amplitude of LH secretory episodes, and rising basal levels of FSH, lead in turn to testicular development involving both spermatogenesis and testosterone production (Fig. 4.2).

In considering the details of the hormonal control of the testes, it is simpler to consider each compartment and its function separately, but we should keep in mind that the two processes of spermatogenesis and steroidogenesis are closely linked.

Fig. 4.2. Changes in the levels of FSH and LH in plasma of adult Soay rams exposed to a controlled lighting regime. Previous exposure to long day-lengths (16 h light:8 h darkness) for 16 weeks had resulted in regression of the testis at the start of the experiment (Day 0). An abrupt change to short day-lengths (8 h light:16 h darkness) resulted in redevelopment of the testes. Outlines at top represent testis growth. Jugular venous blood was collected at hourly intervals on the days indicated. (From G. A. Lincoln and M. J. Peet. *J. Endocrinol.* **74**, 355–67 (1977).)

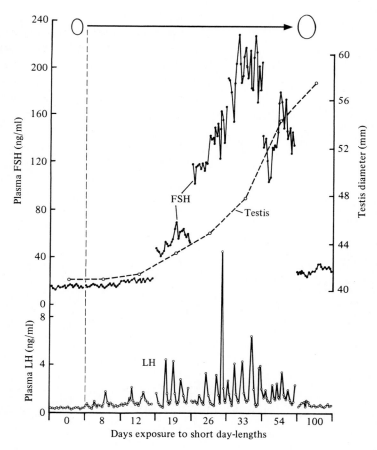

Hormonal control of steroidogenesis

The testis secretes a variety of steroids which are synthesized from cholesterol. The principal secretory product is testosterone, a product of the Leydig cells, which are found in clumps in the intertubular tissue adjacent to the seminiferous tubules (Fig. 4.3). Testosterone is classified as an androgen since it stimulates male secondary sexual characteristics. The synthesis of testosterone proceeds through a biosynthetic pathway, part of which is common to all the major steroid-secreting endocrine glands, the final end product being determined by the enzymatic composition of the tissue. In this pathway, cholesterol is converted to pregnenolone by removal of the C_{21} side-chain, and then through progesterone to several androgenic substances such as dehydroepiandrosterone, androstenedione and testosterone (Fig. 4.4). Although the testes secrete smaller amounts of androstenedione and dehydroepiandrosterone, these are extremely weak androgens and masculinization is due principally to testosterone or its target organ metabolite 5α-dihydrotestosterone.

Fig. 4.3. Diagrammatic representation of the relationships between the seminiferous tubules and interstitial cells. ABP, androgen-binding protein; T, testosterone; E_2, oestradiol-17β.

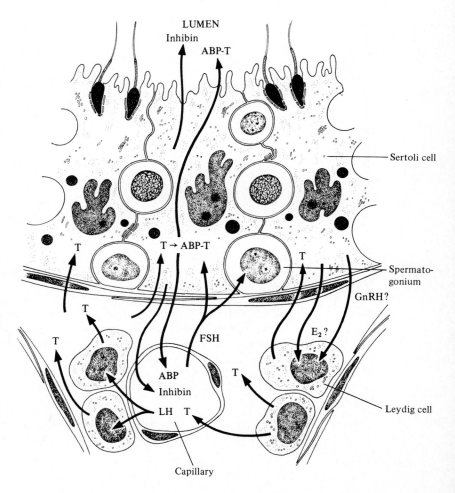

Testosterone secretion by the Leydig cells is stimulated by LH. Receptors for LH are found on the Leydig cells, and in the majority of mammals a rise in LH secretion is followed by a rise in testosterone. In fact, the secretion of both LH and testosterone is episodic, and hence quite large changes in the levels of these two hormones may be found over a 24-h period (Figs. 4.2 and 4.5). The response of the Leydig cells to a rise in LH is quite rapid, peak testosterone levels occurring within 1–2 h though in man very large doses of LH or hCG must be given to demonstrate this acute response. It is important to recognize that LH also has a trophic action on the Leydig cells, stimulating them to undergo hypertrophy. The removal of LH by hypophysectomy or neutralization of its activity by use of a specific antiserum leads to a cessation of testosterone production and a great diminution in the size of the Leydig cells.

The action of LH is mediated through the intracellular formation of 3′–5′ adenosine monophosphate (cyclic AMP) which in turn stimulates, through a protein kinase mechanism, the activation of numerous cellular reactions, one of which is testosterone secretion (for detailed consideration, see Book

Fig. 4.4. Outline of the steroid biosynthetic pathways within the testis.

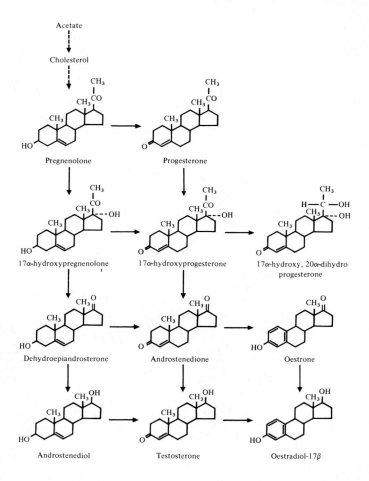

7, Chapter 2. The enzymes necessary for testosterone production are associated with the mitochondria and smooth endoplasmic reticulum of the Leydig cell. Consequently, long-term LH stimulation results in enlargement of the cell together with increases in mitochondria and smooth endoplasmic reticulum (Fig. 4.6). Little is known of the way in which testosterone leaves the Leydig cell but it is found in high concentration in spermatic vein blood, testicular lymph, and in the fluid within the seminiferous tubules.

High doses of LH or hCG were shown by Richard Sharpe and Kevin Catt to cause a large decrease in the number of LH/hCG receptors on Leydig cells for 6–7 days after a single injection. This was accompanied by a decrease in cyclic AMP production and a steep decline in testosterone production, due to an inhibition of certain steps in steroid biosynthesis. We are not clear how this inhibition is mediated, some investigators

Fig. 4.5. Fluctuations in the levels of testosterone (T) and LH in the jugular venous blood of an adult ram. Note the close temporal association between the peaks of LH and testosterone.

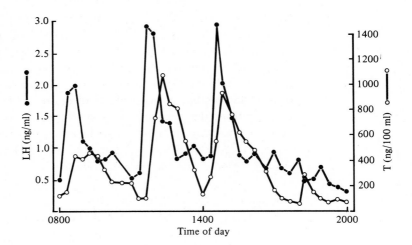

Fig. 4.6. The cytological changes in Leydig cells following chronic LH stimulation. Note the increase in size of the cells as well as the increase in quantity of smooth endoplasmic reticulum and Golgi complex.

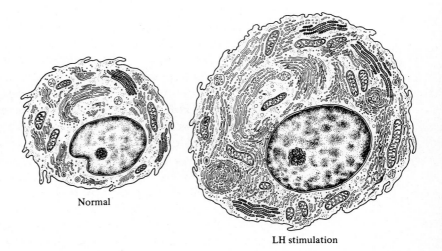

Normal

LH stimulation

suggesting that it is due to increased levels of testicular oestradiol while others proposing that it is due to production of GnRH in the testis (see page 83). While these changes are dramatic, their physiological or pharmacological importance is obscure since our own studies have shown that repeated injections of hCG overcome the block in steroidogenesis, despite continued suppression of LH/hCG receptors to very low levels. It is possible that the trophic action of LH stimulates the growth of existing Leydig cells and the recruitment of immature Leydig cells into the response.

Although LH is the principal factor controlling testosterone secretion, recent evidence suggests that prolactin may also influence the function of the Leydig cell. Receptors for prolactin can be found on Leydig cells, and increased prolactin secretion in men with pituitary tumours is associated with decreased testosterone levels accompanied by diminished libido and the inability to achieve and maintain normal erections. Receptors for GnRH and oestradiol are also present in Leydig cells but the role of these substances in the physiology of the Leydig cell is still unclear.

In many mammalian species, the testis also secretes oestrogen, and in some, such as the stallion and boar, very large quantities are produced. Some of the circulating oestradiol is derived from the peripheral conversion of testosterone, but in man studies of oestradiol levels in spermatic vein blood indicate that approximately 40–50 per cent is secreted by the testis. Some controversy exists as to whether the Leydig cells are the only source of oestradiol since studies in immature rats have shown that the Sertoli cell can convert androgens to oestradiol.

Hormonal control of spermatogenesis

There is general agreement that the induction of spermatogenesis during puberty requires both FSH and LH. The action of LH is indirect, through its stimulation of testosterone production by the Leydig cells, since high intratesticular concentrations of testosterone are required for successful spermatogenesis. Testosterone acts by stimulating the Sertoli cells since the germ cells do not appear to have receptors for testosterone.

Whether FSH is required for the continued maintenance of spermatogenesis once it is successfully established is still controversial and species differences may exist. In hypophysectomized rats, spermatogenesis can continue without FSH, if high doses of testosterone are given immediately after surgery. However, in recent experiments, Raghu Moudgal and Ebo Nieschlag have independently shown that FSH may be required for the maintenance of spermatogenesis in macaque monkeys. They neutralized FSH by passive immunization using highly specific antisera and showed that spermatogenesis and fertility were slightly impaired.

Receptors for FSH are located on the Sertoli cells and spermatogonia, and this hormone exerts its action via cyclic AMP. Injections of FSH are associated with a stimulation of protein kinase activity and protein synthesis. The specialized cell junctions between adjacent Sertoli cells create a barrier to the transport of substances within the intercellular space and effectively isolate the centrally placed germ cells (spermatocytes, spermatids and spermatozoa) from the external environment (Fig. 4.3). Sertoli cells thus support and 'nurse' the germ cells other than the spermatogonia and control their environment (see Book 1, Chapter 4, Second Edition). In fact it seems likely that FSH and LH are involved in stimulating the formation of the specialized cell junctions between adjacent Sertoli cells during pubertal maturation. Use of tissue culture techniques has enabled Sertoli cells to be studied in isolation, and many aspects of their function (Table 4.1) are stimulated by FSH and testosterone which appear to act synergistically.

Although our knowledge of the Sertoli cell has improved considerably over the past decade, we still do not fully understand how it stimulates spermatogenesis. It is important to recognize that the structure of the Sertoli cell varies according to the germ cells surrounding it throughout the stages of the spermatogenic cycle (see Book 1, Chapter 4, Second Edition). Using a transillumination technique, Marti Parvinen has been able to identify and dissect segments of unfixed seminiferous tubules at

Table 4.1. *Effects of FSH and testosterone on the secretory function of Sertoli cells and on the activity of enzymes they contain*

Secretion/activity	Stimulated by:	
	FSH	T
Fluid	+	?
Androgen binding protein	+	+
Inhibin	+	+
Aromatase	+	−
Plasminogen activator	+	?

Table 4.2. *Duration of spermatogenesis in different species*

Species	Duration in days
Boar	34
Mouse	35
Hamster	35
Rat	48–52
Rabbit	49
Ram	49
Bull	54
Man	64

(Data from Y. Clermont, *Physiol. Rev.* **52**, 198–236 (1972).)

differing stages of the seminiferous cycle. This has enabled studies of Sertoli cell function at differing cycle stages, which clearly show that the responsiveness to FSH varies. It seems clear that the function of this cell is complex, but its intimate association with the germ cells makes our knowledge of its function a key requirement to the understanding of spermatogenesis.

We do know that for each species, the time taken for spermatogonia to proceed to spermatozoa is remarkably constant. In the rat this is 48–52 days, whereas in man it is 64 days (Table 4.2). Since the spermatogenic cycle cannot be speeded up, how does FSH stimulate spermatogenesis? Though not conclusive, the results of experiments by Tony Means and Claire Huckins suggest that FSH decreases the percentage of germ cells that normally degenerate during spermatogenesis. This action effectively increases the number of germ cells proceeding through to spermatozoa, hence increasing sperm production.

Interactions between the seminiferous tubules and Leydig cells

Most people accept that the juxtaposition of the Leydig cells with the seminiferous tubules ensures that a high concentration of testosterone continually bathes the seminiferous tubules (Fig. 4.3). This high concentration of testosterone is vital for spermatogenesis, since it stimulates Sertoli cell function and ensures transport of the hormone into the lumen of the seminiferous tubules. The high concentration of testosterone in seminiferous tubule and rete testis fluid is likely to be of importance in maintaining the function of the epididymis.

The concept that the seminiferous tubules influence Leydig cell function is relatively novel. Evidence for such a relationship has been derived by inhibiting spermatogenesis in a variety of ways and demonstrating that there is a resultant hypertrophy of Leydig cells. The behaviour of these cells cannot be due to a circulating humoral factor such as LH, since interfering with spermatogenesis in one testis results in Leydig cell changes in the damaged testis but not in the contralateral normal testis. A variety of factors such as cryptorchidism, irradiation and vitamin A deficiency all interfere with spermatogenesis, and induce Leydig cell hypertrophy and hyperresponsiveness to hCG stimulation. Evidently the altered tubule function has removed an inhibitory influence from the Leydig cells, allowing them to enlarge and become more responsive to trophic stimulation. The source and nature of this inhibitory influence is unknown. Interestingly, in all cases in which spermatogenesis is disrupted, careful studies have revealed that a number of parameters of Sertoli cell function are impaired, suggesting that these cells are a possible source of the signal. Substances such as oestradiol, which are said to be produced by the Sertoli cell, have been suggested as the inter-compartmental regulator. More recently the claim has been made that the Sertoli cells produce a GnRH-like material, and since the Leydig cells have receptors for GnRH this substance could be the signal between the two compartments. This view is supported by the fact that testosterone production by Leydig cells is inhibited by treatment *in vivo* with potent agonists of GnRH. Hence interference with Sertoli cell function could lead to a decrease in GnRH production allowing the Leydig cells to become hyperresponsive. Regardless of the nature of the signal, the postulate that local modulation of Leydig cells may occur requires investigators to consider the testis as an entire functional unit and not as two independent compartments.

The influence of the testes on the pituitary and hypothalamus

The majority of endocrine systems have negative feedback controls whereby the product can regulate the stimulator. That the testis exerts a negative feedback action on the pituitary can be seen by the rise in FSH and LH that follows castration (Fig. 4.7).

The inhibitory action of testosterone on LH secretion has been clearly documented in many species, including man (Fig. 4.8). However, some investigators maintain that it is through conversion to oestradiol that testosterone exerts its negative feedback on LH secretion, and active or

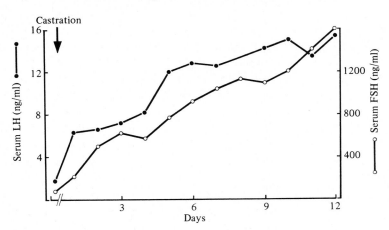

Fig. 4.7. The rise in FSH and LH levels in the serum of rats after castration.

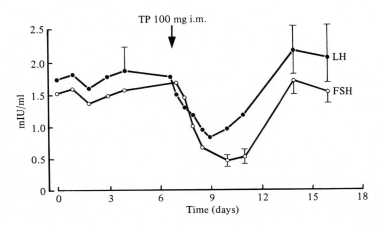

Fig. 4.8. The decline in FSH and LH levels following the intramuscular administration of testosterone propionate (TP) to five normal men. (From H. W. G. Baker. 'The Endocrinology of Liver Disease'. Unpublished Ph.D. thesis, Monash University (1974).)

passive immunization of rams against oestradiol will cause testicular hypertrophy and elevate testosterone levels. Testosterone must be aromatized to oestradiol, either peripherally or in the hypothalamus or pituitary, for this inhibition of LH to occur. However, non-aromatizable androgens such as dihydrotestosterone can also suppress LH. Furthermore, studies by Dick Santen showed that oestradiol and testosterone exert different effects on LH; oestradiol decreases the amplitude but not the frequency of LH pulses, whereas testosterone increases the amplitude but decreases the frequency (Fig. 4.9). These distinctions suggest that both oestrogen and testosterone may be important in controlling LH secretion, with oestradiol acting at the pituitary and testosterone altering GnRH secretion by the hypothalamus.

The availability of pituitary cell cultures has shown that oestradiol and testosterone can exert their actions directly on the pituitary. However, they may also modify the secretion of GnRH by the hypothalamus. Since the pulsatility of LH secretion is highly likely to be the result of episodic GnRH secretion, the effect of testosterone on the frequency of LH pulses described above (Fig. 4.9) is most probably mediated by an action on the hypothalamus.

Testosterone and oestradiol also inhibit FSH secretion, and this observation has led to the postulate of a common negative feedback influence for both gonadotrophic hormones. However, there are some instances where FSH levels rise disproportionately to LH levels, such as following the induction of spermatogenic damage in rodents or in states of severe spermatogenic disruption in infertile men (Fig. 4.10). Such observations

Fig. 4.9. The effects of infusions of testosterone and oestradiol on LH pulse frequency in men. The solid line represents an idealized pattern. Note the increase in pulse amplitude and decrease in pulse frequency during the testosterone infusion, and the decreased pulse amplitude and unchanged frequency during the oestradiol infusion. The numbers below the recordings indicate the pulse frequency per 6 h. (From R. J. Santen. In *The Testis in Normal and Infertile Men*, ed. P. Troen and H. R. Nankin. Raven Press; New York (1977).)

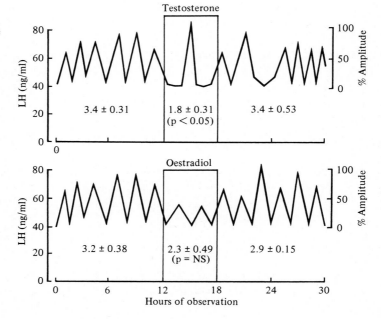

have led to the concept that there may be a specific factor inhibiting FSH secretion. The substance was termed inhibin by David McCullagh in 1932, but its complete characterization has still not been achieved. Testicular extracts, rete testis fluid, testicular lymph, semen, and ovarian follicular fluid contain a proteinaceous material capable of selectively lowering FSH secretion both *in vivo* and *in vitro*. Media from cultures of Sertoli cells have been shown to contain inhibin, but as yet inhibin-like activity has not been detected in the blood of males. Very recent studies in the female have shown that inhibin levels in blood are increased following gonadotrophic stimulation of the ovary.

The secretion of FSH by dispersed pituitary cells in culture can be used as a bioassay for inhibin, so it is clear that inhibin can act directly on the pituitary. Whether inhibin also exerts an inhibitory action at the

Fig. 4.10. Serum FSH and LH levels in infertile men with the Sertoli-cell-only syndrome (germ cell aplasia). The shaded area indicates the normal range. Note the larger increase in FSH than LH.

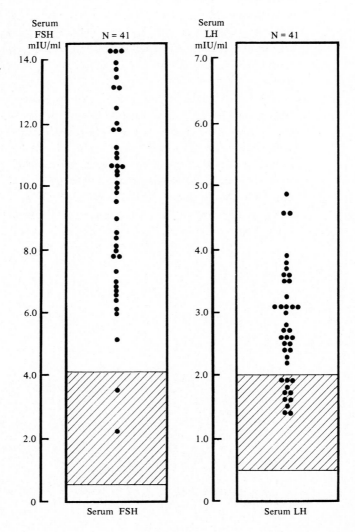

hypothalamus is still uncertain. In the pituitary, inhibin selectively suppresses the synthesis and release of FSH, both basally and following GnRH stimulation; it only causes LH suppression when administered in high doses. Consequently, the likelihood is that steroids and inhibin can both act directly on the pituitary to influence FSH and LH secretion. Since there is only one releasing hormone for FSH and LH, and since 90 per cent of the pituitary gonadotrophs secrete both gonadotrophins (which share a common α-subunit), the intracellular mechanisms governing the ratio of FSH and LH secreted must be particularly complex. At the moment, little is known of how this regulation is achieved.

Endocrine changes in testicular disorders associated with male infertility
Our understanding of the endocrine control of testicular function combined with the development of sensitive and specific methods of FSH and LH measurement has been of value in improving the management of male infertility. Estimates suggest that approximately one man in 25 is subfertile, but careful hormonal studies indicate that less than one per cent of these cases is due to decreased gonadotrophin stimulation. This figure is in strong contrast to the female where disorders of gonadotrophin stimulation of the ovary account for approximately 40–50 per cent of the cases of infertility. Nevertheless measurements of the gonadotrophic hormones, particularly FSH, have proved extremely valuable in the diagnosis of certain testicular disorders.

By extensive studies in which we compared the histological appearance of the testis with the levels of FSH in the serum of men with lowered sperm counts, we were able to demonstrate a negative correlation between FSH levels and the severity of the seminiferous tubule damage (Fig. 4.11). Thus, men in whom the seminiferous epithelium was severely disrupted, such as

Fig. 4.11. Serum FSH levels from men with testicular disorders presenting as infertility. The patients have been divided into categories on the basis of their testicular histology. The shaded areas indicate the normal ranges. (From H. W. G. Baker *et al.* (1976) – see Suggested Further Reading.)

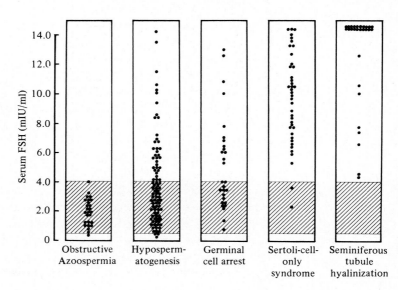

those with biopsy categories of seminiferous tubule hyalinization and Sertoli-cell-only syndrome (Fig. 4.12*b* and *c*), showed greatly elevated levels of serum FSH, presumably due to lack of inhibin production by the testis. Conversely, in men whose sperm count was zero (azoospermia) but in whom the testis biopsy was normal (Fig. 4.12*a*), serum FSH levels were also within the normal range, the absence of sperm in the ejaculate being due to an obstruction within the rete testis, epididymis or vas deferens. Since men with obstruction and those with severe seminiferous tubule damage all present clinically with azoospermia, measurement of FSH permits differentiation between the two groups on the basis of a blood sample. Those with obstruction have normal levels of FSH and require reconstructive surgery, whereas those with seminiferous tubule damage have elevated FSH levels and irreversible sterility since no germ cells are present in the testis. With less severe disruption of spermatogenesis, for instance in men with hypospermatogenesis (Fig. 4.11), the finding of an elevated level of FSH usually indicates that treatment is unlikely to result in a stimulation of sperm count.

The assessment of LH and testosterone levels in these infertile males has also shown that severe destruction of the seminiferous tubules, as in hyalinization or the Sertoli-cell-only syndrome, leads to elevated LH levels and low to normal or frankly low testosterone levels (Fig. 4.10). These observations indicate some degree of Leydig cell failure, since higher than normal LH levels are required to maintain the low–normal testosterone levels. Stimulation with hCG often leads to a subnormal response of testosterone and some men benefit from androgen replacement therapy (see Book 8, Chapter 2, First Edition). Perhaps the most severe form of testicular damage is seen in men with Klinefelter's syndrome (see Book 2, Chapter 3, Second Edition), who most often require testosterone treatment for symptoms such as loss of libido and potency, tiredness, hot flushes and poor body-hair growth. The detection of androgen deficiency is important and may require more than one blood sample, in view of episodic secretion, or the use of hCG as a provocative stimulus capable of detecting a decrease in the capacity of the testis for testosterone production.

We do not yet know whether changes in the tubules alter Leydig cell function in man, although the fact that Leydig cell activity is impaired in men with severe seminiferous tubule damage suggests that this may be the case. Much remains to be learnt about the factors that initiate and maintain spermatogenic arrest in such disorders.

In the First Edition of this book, written 11 years ago, the testis received scant attention since at that time so little was known about it. There has recently been a burgeoning interest in testicular physiology and endocrinology, and we are now beginning to be able to listen in to the fascinating cross-talk that goes on between the germ cells, the Sertoli cells and the Leydig cells. Unfortunately we are still frustrated in our attempts

Fig. 4.12. The photomicrographs indicate the different appearances of testicular biopsies from: (*a*) A man with obstruction and a normal biopsy. (*b*) A man with Sertoli-cell-only syndrome. (*c*) A man with seminiferous tubule hyalinization.

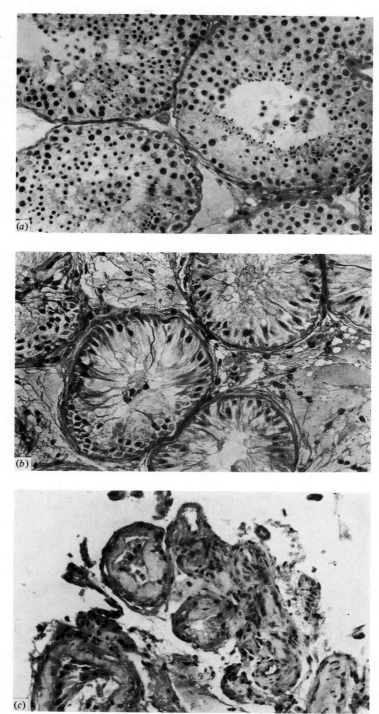

to treat most forms of male infertility, and it is also proving particularly difficult to develop new approaches to male contraception. In the latter case, since a man's sexual behaviour is in part dependent on his testosterone secretion, we are forced to search for compounds that will leave his androgen levels undisturbed, whilst totally suppressing either the production or the function of up to 100 million spermatozoa per day – no easy task. Inhibin once seemed like a promising candidate, since it offered the possibility of regulating FSH secretion independently of LH. Unfortunately spermatogenesis, once established, seems to be relatively independent of FSH, and inhibin would in any case be difficult to administer on a long-term basis since it is probably a polypeptide. Perhaps we must just accept the fact that we need to learn a great deal more about testicular physiology before we can hope to be able to manipulate male fertility successfully.

Suggested further reading

Seasonal breeding: nature's contraceptive. G. A. Lincoln and R. V. Short. *Recent Progress in Hormone Research*, **36**, 1–43 (1980).

Is aromatization of testosterone to estradiol required for inhibition of luteinizing hormone secretion in men? R. J. Santen. *Journal of Clinical Investigation*, **56**, 1555–63 (1975).

The hormonal regulation of the Leydig cell. R. M. Sharpe. In *Oxford Reviews of Reproductive Biology*, vol. 4, pp. 241–317. Ed. C. A. Finn. Clarendon Press; Oxford (1982).

Functional relationships of the mammalian testes and epididymis. G. M. H. Waites. *Australian Journal of Biological Sciences*, **33**, 355–370 (1980).

Testicular control of follicle-stimulating hormone secretion. H. W. G. Baker, W. J. Bremner, H. G. Burger, D. M. de Kretser, A. Dulmanis, L. W. Eddie, B. Hudson, E. J. Keogh, V. W. L. Lee and G. C. Rennie. *Recent Progress in Hormone Research*, **32**, 429–69 (1976).

The Testis. Ed. H. G. Burger and D. M. de Kretser. Raven Press; New York (1981).

5

The ovary

DAVID T. BAIRD

Unlike the testes, the ovaries of all mammals remain within the abdominal cavity where they are well protected from injury by external agents. The ovary is subdivided into a series of specialized compartments or structures, each with its own precisely regulated micro-environment. In this way the oocytes can be nursed through from the start of oogenesis until ovulation (see Book 1, Chapters 1 and 2, Second Edition). The endocrine function of the ovary ensures the regular production of healthy oocytes at a time when they will have a maximum chance of being fertilized. Hence, although it is often convenient to consider the oogenic and endocrine functions of the ovary separately, the two are intimately interconnected. An ovary devoid of oocytes cannot function normally as an endocrine gland.

In the mature animal the structure and function of the ovary is continually changing. Gonadotrophins secreted by the anterior pituitary gland stimulate the growth of Graafian follicles (folliculogenesis), ovulation, and the formation of corpora lutea. The time taken for follicles and corpora lutea to develop differs from species to species and is reflected in different patterns of ovarian cycles. Hence while ovarian function is closely regulated by a feedback system involving the hypothalamus and anterior pituitary, the 'zeitgeber' or biological clock which determines the length of the cycle is the ovary itself.

Morphology of the ovary

The ovaries are paired organs situated within the abdominal cavity and covered by a single layer of surface epithelium, formerly misleadingly called the germinal epithelium, which is continuous with the lining of the peritoneal cavity. The integrity of this covering is periodically broken at ovulation so that the surface epithelium is in a continuous state of damage, repair and regeneration. In a mature woman each ovary weighs about 8 g, but because much of it is made up of transitory structures such as follicles and corpora lutea, its relative size changes during the cycle.

The ovary is well supplied with blood by an ovarian artery and numerous anastomosing branches from the uterine artery (Fig. 5.1). In most species the arterial supply divides into several smaller coiled branches close to the ovary, which provide an efficient means for adapting to changes in structure. The necessity for rapid change is well illustrated by the vascularization of the corpus luteum. Within a few days of ovulation a whole new network of capillaries and arterioles with their accompanying venous supply is established. The mature corpus luteum has one of the richest blood supplies per gram of any tissue in the body, including the renal cortex.

The venous drainage from the ovary is by a network of venules which surround the arteries. In many species the ovarian and uterine venous blood drains into a common utero-ovarian vein on which branches of the ovarian artery wind (Fig. 5.2). The close association of ovarian artery with utero-ovarian vein provides an anatomical basis for the transfer of hormones, such as prostaglandins and steroids, from vein to artery. This

Fig. 5.1. The blood supply to the human ovary and uterus. Note the extensive anastomoses between the ovarian and uterine vessels. (From R. De Graaf. De Mulierum Organis Generationi Inservientibus, Leyden (1672).)

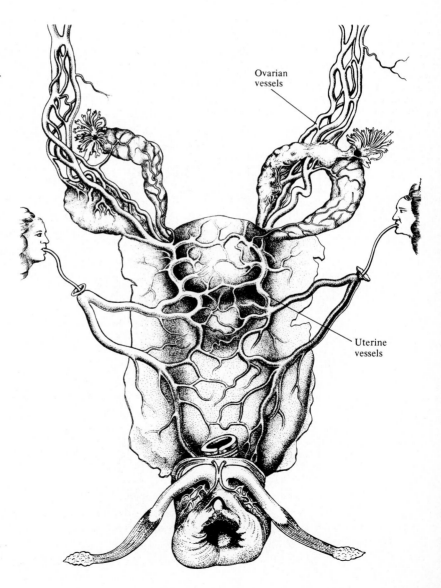

Ovarian
vessels

Uterine
vessels

countercurrent transfer is a mechanism whereby products of the uterus or ovary can feed back locally to influence ovarian function. It is the means by which the uterus in many species induces regression of the corpus luteum (see below).

Accompanying the blood vessels is a rich network of lymphatics and nerves. Both parasympathetic and sympathetic nerves terminate in the ovary, but their importance in regulating ovarian function is not known. In the dog at any rate they do not appear to be essential, because ovulation and pregnancy have occurred in bitches after total denervation of the ovaries.

Fig. 5.2. The blood supply to the ovary and uterus in the sheep. The blood vessels were injected with latex and the tissues fixed and cleared. Note the convolutions of the ovarian artery, which is closely applied to the ovarian and utero-ovarian vein. (From C. H. Del Campo and O. J. Ginther. *Amer. J. Vet. Res.* **34**, 305–16 (1973).)

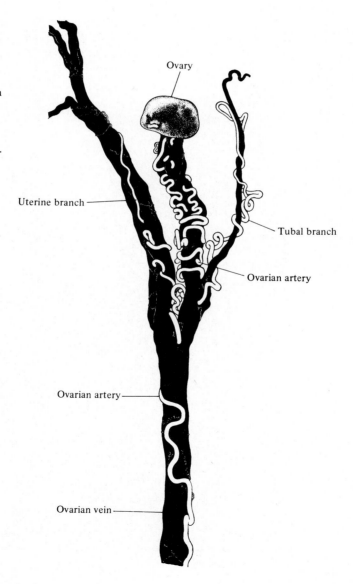

It is convenient to consider the adult ovary as composed of three separate functional units or compartments, the stroma and interstitial tissue, the follicles, and the corpus luteum. It should be realized, however, that these structures are in a constant state of flux and as one regresses another may be formed from its elements – the follicle is converted into the corpus luteum, and atretic follicles contribute to the stroma.

Stroma

The stroma is derived from coelomic mesenchyme and provides the basic cellular matrix in which the follicles are distributed. Scattered throughout the cortex of the ovary are many thousands of primordial follicles formed from the residue of the stock of oocytes present in fetal life. It is not yet clear how individual oocytes are recruited to join the pool of growing follicles, but once growth is initiated they induce changes in the surrounding stromal cells which then differentiate into granulosa and thecal cells. Conversely when follicles degenerate and become atretic, many of the associated cells de-differentiate to stromal cells. In some species such as the rabbit and hare these cells are sufficiently prominent to be called

Fig. 5.3. Interstitial gland of the snowshoe hare (*Lepus americanus*) in early pregnancy showing corpus luteum (left). (From H. W. Mossmann and K. L. Duke. *Comparative Morphology of the Mammalian Ovary*. University of Wisconsin Press (1973).)

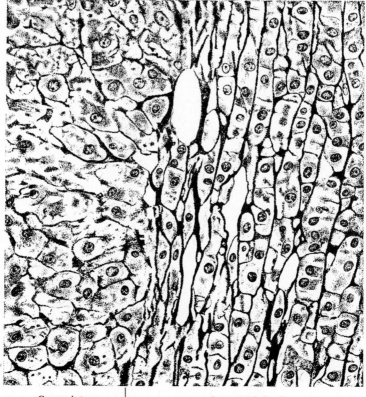

Corpus luteum | Interstitial gland

an interstitial gland, which secretes large amounts of steroid hormones (Fig. 5.3).

In the human and certain other mammals, groups of large polyhedral cells can be identified near the hilus of the ovary at the point of entry of the ovarian blood vessels. These cells are morphologically similar to the Leydig cells of the testis and are thought to produce androgens. Rarely these cells may form tumours (hilar cell tumours) which cause masculinization through the secretion of testosterone.

Follicle

Follicles at all stages of development can be found distributed throughout both ovaries at all times, except after the menopause in women. The details of folliculogenesis have been described in the Second Edition of Book 1, Chapter 2, and here we need only consider the anatomy of the larger follicles. The mature Graafian follicle is composed of several layers of cells surrounding the oocyte, which is contained within a fluid-filled cavity or antrum (Fig. 5.4). The outermost layers, the theca externa and theca interna, are formed from the adjacent stromal cells. They are supplied by a rich network of capillaries and separated by a basement membrane from the avascular granulosa cells which line the follicular cavity. The granulosa cells are comparable to the Sertoli cells of the testis and are important in maintaining the very specialized conditions within the antral cavity which

Fig. 5.4. The structure of the Graafian follicle shows how the granulosa cells are deprived of a blood supply by the basement membrane.

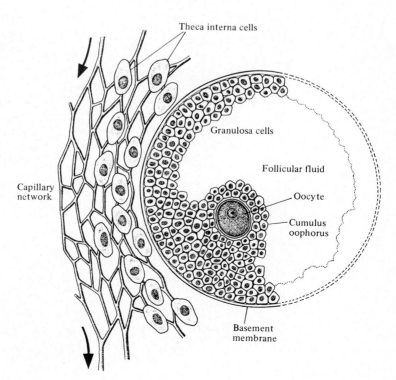

permit development of the oocyte. The population of granulosa cells is probably not homogeneous, since those lining the basement membrane and those surrounding the oocyte (the cumulus oophorus) serve different functions; for example, there are microtubular connections between the cumulus and oocyte which are probably important for transferring nutrients and hormones to the oocyte.

Corpus luteum

The corpus luteum is formed from the granulosa and thecal cells of the follicle ruptured at ovulation. The number of corpora lutea present in the ovaries is directly related to the number of eggs shed, and varies with the species (Fig. 5.5). In monovular species, such as the human, ovulation occurs at random in either ovary. The newly formed corpus luteum rapidly becomes invaded by blood vessels, and these follow strands of thecal cells interspersed between the larger granulosa luteal cells (Fig. 5.6). There are at least two types of luteal cell, large and small, which are thought to originate from the granulosa and thecal cells of the follicle respectively. In some species, such as the human, a third type of luteal cell (K cell) can be identified histologically, although its function is not known.

Fig. 5.5. Ovaries of a woman (*a*) and plains viscacha (*b*) in early pregnancy showing the marked variation between mammals in the number of follicles ovulated. (*a* from H. W. Mossman and K. L. Duke. *Comparative Morphology of the Mammalian Ovary*, University of Wisconsin Press (1973); *b* from a photograph, by courtesy of Dr Barbara Weir.)

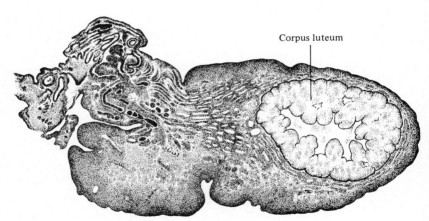

Corpus luteum

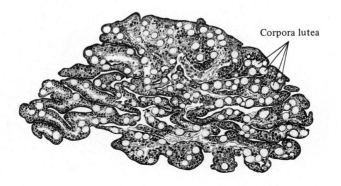

Corpora lutea

The increase in weight of the corpus luteum after ovulation is largely due to an increase in the size of the luteal cells (hypertrophy) rather than an increase in their number (hyperplasia). Thus the number of cells within the corpus luteum is similar to that found in the pre-ovulatory follicle.

Hormones secreted by the ovary

It has been known for many years that the ovary secretes a variety of steroid hormones, including oestradiol-17β and progesterone. Oestradiol-17β was first extracted from pig follicular fluid by Edgar Allen and Edward Doisy at Washington University, St. Louis, in 1923. More recently it has been discovered that the ovary also synthesizes a great variety of non-steroidal substances with hormonal actions, i.e. they influence the structure or function of other cells. The chemical structure of some of these other ovarian hormones is known, for example the prostaglandins, relaxin, and oxytocin: the presence of others is inferred from their biological activity, as in the case of inhibin. While some of these ovarian hormones are secreted into the bloodstream via the ovarian venous or lymphatic system, others exert their main action locally within the ovary.

Steroids

The most important steroid hormones secreted by the ovary are progesterone and oestradiol-17β. All steroids are synthesized from cholesterol, which is produced from acetate within the cell, or transported in the blood plasma bound to low-density lipoprotein (LDL) (Figs. 5.7 and 5.8). The biosynthetic steps from cholesterol to pregnenolone are common for all steroid hormones including the corticosteroids.

Fig. 5.6. The corpus luteum of a woman on day 20 of the cycle. Note the strands of thecal cells (T) between the granulosa cells (G). bv, blood vessel. (From D. T. Baird, T. G. Baker, K. P. McNatty and P. Neal. *J. Reprod. Fert.* **45**, 611–19 (1975).)

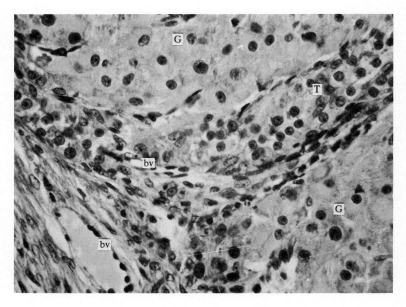

Figure 5.8 represents the main route of biosynthesis of steroid hormones by the ovary. The oxidation of pregnenolone to form progesterone is catalaysed by the enzyme 3β-hydroxysteroid dehydrogenase which is present in the thecal and granulosa cells of the pre-ovulatory follicle as well as in the cells of the corpus luteum. Progesterone is the hormone responsible for the maintenance of pregnancy; it causes relaxation of smooth muscle and reduced excitability of the myometrium. For progesterone to exert its effect, the tissue must first be exposed to oestrogen. Together with oestrogen, progesterone induces development of the lobulo-alveolar system of the breast, and hypertrophy of the uterine endometrium.

Pregnenolone and progesterone are converted to the *androgens* dehydro-epiandrosterone, androstenedione and testosterone by hydroxylation at position 17 followed by removal of the C_{21} side-chain (Fig. 5.9). In some species the biologically 'weak' androgen, androstenedione, is quantitatively the major product secreted by the ovary at the time of ovulation. In the spotted hyaena the ovarian secretion of testosterone is apparently sufficient to produce growth and development of male external genitalia (see Book 2, Chapter 3, Second Edition). Although the function of ovarian androgens in women is not known, it is likely that they play some role in maintaining sexual libido (see Book 4, Chapter 5, in this Edition).

Oestrogens are synthesized from androgens by a series of enzymatic reactions collectively known as 'aromatization'. In every mammalian species studied so far the mature Graafian follicle secretes large amounts of oestradiol-17β. Some years ago Bengt Falck in Sweden performed some classic experiments in the rat involving transplantation of the different cell types of the follicle to the anterior chamber of the eye. By observing the cellular changes in a small piece of vagina placed next to the follicle cells he was able to assess the production of oestrogen. He found that the maximal synthesis of oestrogen occurred when both granulosa and thecal cells were placed near each other and hence concluded that some interaction between the two cell types probably occurred in the follicle *in vivo*. Since

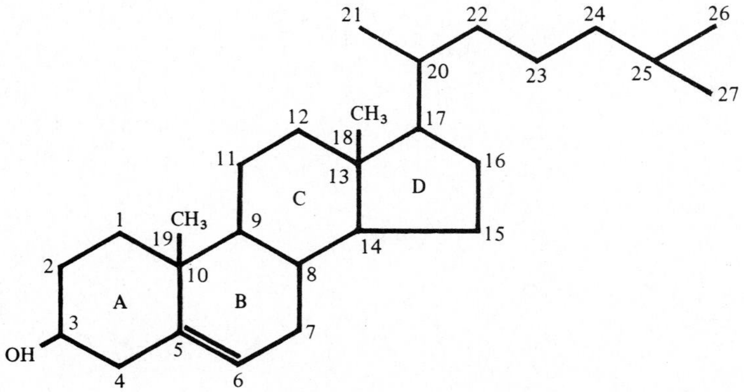

Fig. 5.7. The structure of cholesterol with the numbering of the carbon atoms and identification of the cyclopentanophenanthrene rings.

Fig. 5.8. Biosynthesis of progesterone, androgens and oestrogens from acetate and cholesterol.

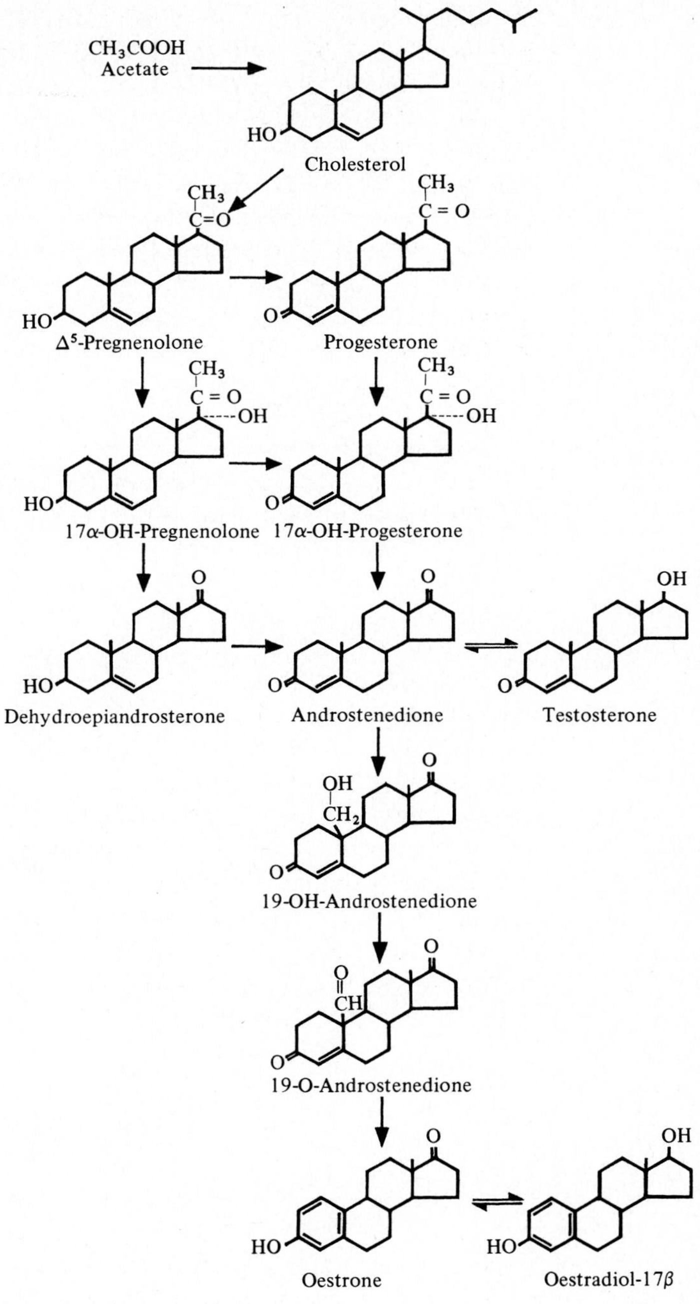

CH₃COOH Acetate →

Cholesterol

Δ⁵-Pregnenolone

Progesterone

17α-OH-Pregnenolone

17α-OH-Progesterone

Dehydroepiandrosterone

Androstenedione

Testosterone

19-OH-Androstenedione

19-O-Androstenedione

Oestrone

Oestradiol-17β

then considerable evidence has accumulated to support the 'two cell' theory for the synthesis of oestrogen by the follicle (Fig. 5.10). Under stimulation by LH, the thecal cells synthesize mainly androgens which are either secreted into the ovarian vein or traverse the basement membrane to reach the interior of the follicle (Fig 5.11). Granulosa cells of mature follicles have a very high aromatase activity and convert most of the androgens produced by the thecal cells into oestrogens. In this way the intrafollicular environment of the follicle is kept highly oestrogenic. The activity of the aromatase enzyme is stimulated by FSH (see below).

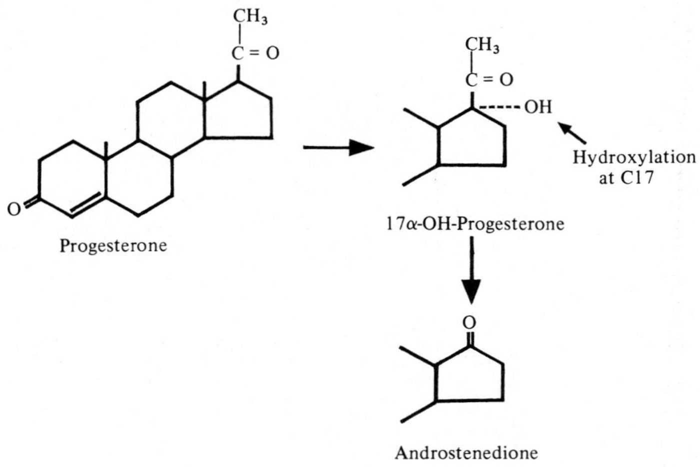

Fig. 5.9. The conversion of a 21-carbon steroid, progesterone, to a 19-carbon androgen, androstenedione, by hydroxylation at position 17 followed by cleavage of the side-chain.

Fig. 5.10. Biosynthesis of steroids by the Graafian follicle. LDL, low-density lipoprotein.

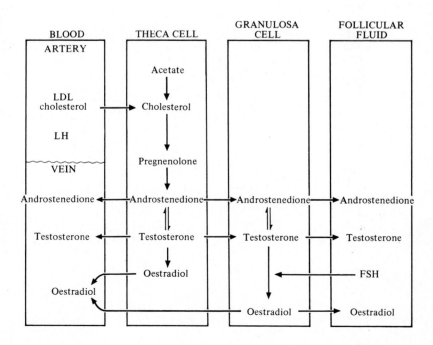

In addition to influencing the secretion of gonadotrophins via their feedback effects on the hypothalamo-pituitary system, as discussed by Fred Karsch in the opening chapter, ovarian steroid hormones also exert important actions locally within the ovary. Oestradiol-17β is present within the antrum of the developing Graafian follicle in a concentration which is 1000-fold higher than that found in the bloodstream. This high local concentration of oestrogen maintains a favourable environment for the growth and development of Graafian follicles. For example, massive growth of granulosa cells can be induced in immature hypophysectomized rats by implanting them subcutaneously with stilboestrol – a synthetic oestrogen (Fig. 5.12). This experimental model has been widely used to

Fig. 5.11. Diagram of action of gonadotrophins on the follicle and the synthesis of oestrogens. LH interacts with receptors on the theca cells (■) to stimulate production of androgens and small amounts of oestradiol. FSH activates the aromatase enzyme system in the granulosa cells by interacting with receptors (□). A, androstenedione; T, testosterone; LH, luteinizing hormone; FSH, follicle stimulating hormone; E$_2$, oestradiol.

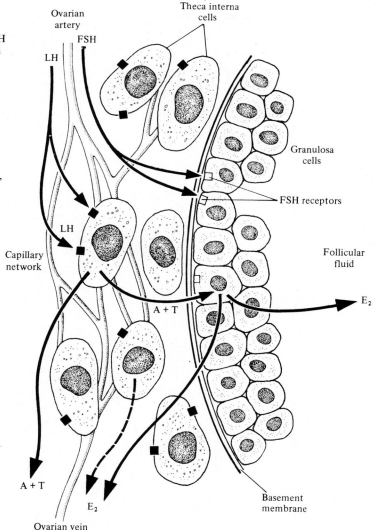

Fig. 5.12. The appearance of ovaries from hypophysectomized rats, untreated (*a*), or treated with 2 mg diethylstilboestrol (DES) (*b*), or with 200 μg FSH, (*c*) or with 2 mg DES followed by 200 μg FSH (*d*). (*b*, *c*, and *d* from R. L. Goldenberg, J. L. Vaitukaitis and G. T. Ross. *Endocrinology*, **90**, 1492–8 (1972).)

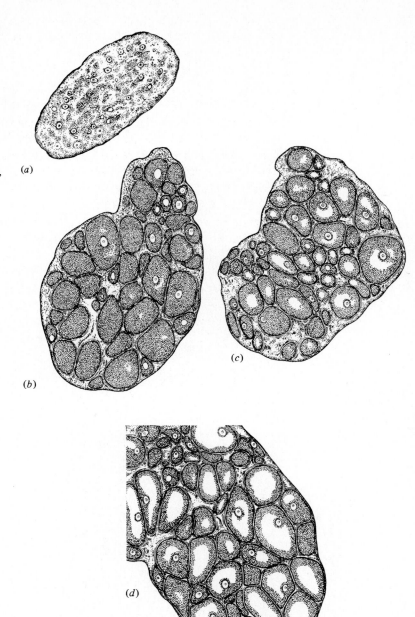

(*a*)

(*b*)

(*c*)

(*d*)

study the factors which influence the differentiation of the granulosa cell and the development of receptors for gonadotrophins.

Non-steroidal hormones

Prostaglandins are a group of lipid-soluble acids which are found in many tissues of the body (Fig. 5.13). The granulosa cells of the pre-ovulatory follicle produce large amounts of prostaglandin E_2 (PGE$_2$) and prostaglandin $F_{2\alpha}$ (PGF$_{2\alpha}$) which are thought to play some role in rupture of the follicle. PGF$_{2\alpha}$ synthesized by the uterus and transported by a counter-current mechanism to the adjacent ovary is the factor responsible for regression of the corpus luteum in many species including the horse, cow, sheep, pig and guinea pig (Fig. 5.14). It has been suggested that a local production of prostaglandins within the corpus luteum is responsible for luteolysis in primates, in which the uterus itself plays no role in luteal regression.

The corpus luteum synthesizes at least two polypeptide hormones – relaxin and oxytocin. The former is produced mainly during pregnancy, when it induces softening of the inter-pubic ligament and reduces the contractility of the uterus (see Chapter 7). It is also found in the corpus luteum of the cycle, although its role is unknown. The structure of relaxin

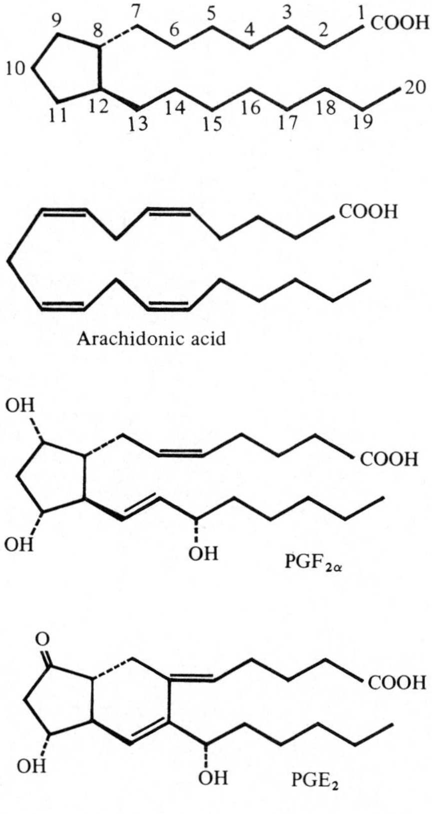

Fig. 5.13. Structure and number of carbon atoms of prostanoid skeleton and structure of arachidonic acid and prostaglandins $F_{2\alpha}$ and E_2.

is very similar to that of insulin, being a basic polypeptide of 48 amino acid residues organized with two chains linked by disulphide bridges through cysteine (Fig. 5.15).

Very recently oxytocin has been found in relatively high concentrations in the corpus luteum of the sheep and human. It is apparently secreted into the ovarian vein in both these species, but its physiological significance is as yet unknown. It may be a local regulator of ovarian function (see Chapter 7).

A variety of other factors are thought to be produced by the follicle at different stages of its development. In the male, the presence of a substance called inhibin, which is not a steroid but which suppresses the secretion of FSH by the pituitary, has been suspected for nearly 50 years (see Chapter 4). Though the molecular structure of inhibin is not yet known, there is convincing evidence that it is a protein synthesized by the Sertoli cells of the testis. More recently large amounts of inhibin-like activity have been found in follicular fluid and in the fluid in which granulosa cells have been cultured, so it seems likely that the ovary does produce at least one protein which modifies the secretion of FSH. Whether inhibin has any local role within the ovary is not yet known.

Follicular fluid also apparently contains a variety of other proteins with biological properties which may be important in regulating intra-ovarian activity. Nina Channing has suggested that arrest of the oocyte in the

Fig. 5.14. Route by which PGF$_{2\alpha}$ produced by the progesterone-primed uterus is able to enter the ovarian artery and induce regression of the corpus luteum in the sheep.

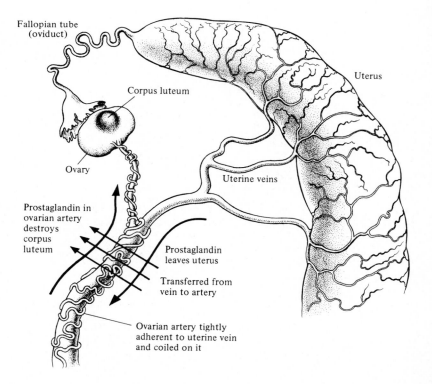

dictyate stage of the first meiotic division is due to the presence of a factor which inhibits meiosis (the oocyte maturation inhibitor or OMI; c.f. Anne Grete Byskov's meiosis-preventing substance or MPS, discussed in Book 1, Chapter 1, in this Edition). Peptides similar in structure to gonadotrophin releasing hormone (GnRH) have been isolated from the testis, and GnRH has a direct effect on the synthesis of steroids by the ovary. Although such a peptide has not yet been isolated from the ovary, the presence of specific ovarian receptors for GnRH, very similar to those found in the anterior pituitary, suggests that there may be some physiological role for this compound in the ovary.

Gonadotrophins and the ovary

We have known for over 50 years that the functioning of the ovary is totally dependent on the secretion of gonadotrophins from the anterior pituitary. Steroids and possibly other factors secreted by the ovary in turn feed back onto the hypothalamus and pituitary to regulate the secretion of gonadotrophins. However, the response of the ovary to gonadotrophins depends on the presence of specific receptors on the surface of the different ovarian cell types. Although these receptors are specific for individual gonadotrophins, e.g. LH will only interact with its specific receptor, the hormones appear to induce a response in the cell via a common mechanism involving adenylate cyclase, the enzyme responsible for stimulating the production of cAMP from ATP (see Book 7, Chapter 2, First Edition).

Follicle stimulating hormone and luteinizing hormone are both glycoproteins with molecular weights about 30 000. The carbohydrate content (27 per cent for FSH) and in particular the sialic acid content (FSH 5 per cent and LH 1.4 per cent) are essential for their biological action. Both FSH and LH are composed of two subunits, one of which (the α-subunit) is common to both gonadotrophins and to thyroid-stimulating hormone (TSH). This common α-subunit explains the degree of immunological cross-reaction that occurs between these three pituitary hormones. The β-subunit is specific for each hormone and determines the biological activity. FSH promotes the growth and differentiation of follicles in the ovary and is essential for formation of an antrum. Although classically FSH was considered to be devoid of steroidogenic activity, in recent years Jennifer Dorrington and David Armstrong in Canada have discovered that FSH specifically stimulates the conversion of androgens to oestrogens. As

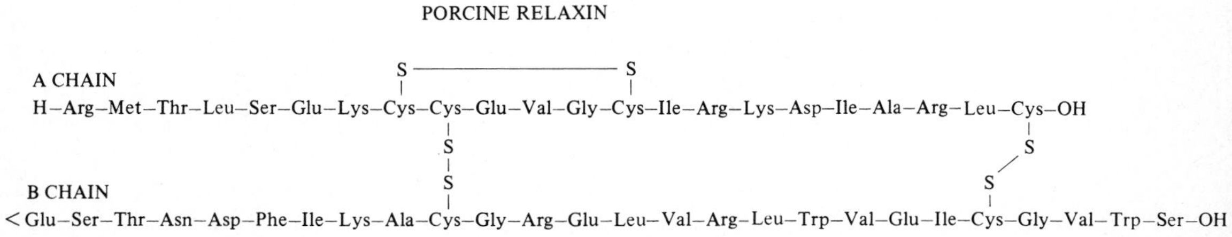

Fig. 5.15. Structure of porcine relaxin.

PORCINE RELAXIN

A CHAIN

H—Arg—Met—Thr—Leu—Ser—Glu—Lys—Cys—Cys—Glu—Val—Gly—Cys—Ile—Arg—Lys—Asp—Ile—Ala—Arg—Leu—Cys—OH

B CHAIN

< Glu—Ser—Thr—Asn—Asp—Phe—Ile—Lys—Ala—Cys—Gly—Arg—Glu—Leu—Val—Arg—Leu—Trp—Val—Glu—Ile—Cys—Gly—Val—Trp—Ser—OH

receptors for FSH are present on the granulosa cells of all healthy follicles, the ability of FSH to stimulate the aromatase enzyme system is probably important in maintaining a high level of oestrogen synthesis within the developing follicle (see Fig. 5.11). It is interesting that the Sertoli cell in the testis also contains receptors for FSH and will convert androgens to oestrogens.

LH has several distinct actions on the ovary. First, following its administration there is a rapid increase in ovarian blood flow which is probably important in maintaining an increased supply of precursors and removing metabolites. The most important action of LH, however, is to increase the synthesis of steroids by those cell types in the ovary which have receptors for LH, i.e. the stroma, theca interna and granulosa cells of pre-ovulatory follicles, and later the corpus luteum; this is done by stimulating the conversion of cholesterol to pregnenolone. The final product secreted depends on the relative activity of the other steroid-converting enzymes. Hence in the corpus luteum abundant 3β-ol-dehydrogenase and little side-chain cleavage results in the production of large quantities of progesterone. In the cells of the stroma and theca interna on the other hand, pregnenolone is converted to androstenedione and testosterone via 17α-hydroxypregnenolone and dehydroepiandrosterone.

FSH and LH are both secreted by the same cell in the anterior pituitary gland in response to stimulation by gonadotrophin releasing hormone (GnRH). The secretion of GnRH by the hypothalamic neurones is intermittent and hence the release of gonadotrophins is pulsatile in nature. Thus the ovary is exposed to a fluctuating rather than a constant concentration of LH and FSH. In those species which have been studied intensively, like the sheep and the human, there is a striking increase in the frequency of LH pulses during the follicular phase of the cycle. Each pulse of LH is followed by a marked increase in the secretion of androgens and oestrogens by the ovary (Fig. 5.16). Thus the distinctive increase in oestrogen secretion from the dominant follicle prior to ovulation is due to stimulation by a series of LH pulses.

In addition to its important action of stimulating steroid synthesis, LH induces ovulation in follicles which have been adequately primed with FSH. The ovulatory action of LH is probably separate from its steroidogenic one, although it will only affect those mature follicles whose granulosa cells have acquired receptors for LH. The surge of pituitary LH responsible for ovulation is produced by the 'positive feedback' action of oestradiol on the hypothalamus and anterior pituitary (see Chapters 1 and 6 in this Book).

Although FSH and LH are normally secreted by the same pituitary cell, the proportions differ depending on the physiological situation; the FSH:LH ratio is in part determined by the pulsatile frequency of GnRH secretion. FSH interacts with oestradiol to stimulate the development of

LH receptors on the granulosa cells of the maturing follicle. Thus the development of the follicle depends on a carefully coordinated sequence of events which ensures that the granulosa and thecal cells as well as the oocyte are fully mature by the time ovulation takes place.

The anterior pituitary secretes a third gonadotrophin – prolactin. Although its exact role in regulating ovarian function is complex and not fully understood, receptors for prolactin have been identified on the granulosa cells of large follicles and on luteal cells of some species. Prolactin interacts with a receptor on the surface membrane of cells but it does not appear to activate the adenylate cyclase enzyme in the same way as do LH and FSH. During the luteal phase of the cycle prolactin together with LH is probably required for the maintenance and proper functioning of the corpus luteum of many species. There is evidence that under certain circumstances prolactin may suppress follicular growth, perhaps by inhibiting the conversion of androgens to oestrogens, and in

Fig. 5.16. Changes in secretion of ovarian and pituitary hormones around ovulation in the sheep. The top half of the diagram illustrates the marked increase in the frequency of LH pulses in the follicular phase of the cycle as the secretion of progesterone declines during luteal regression.

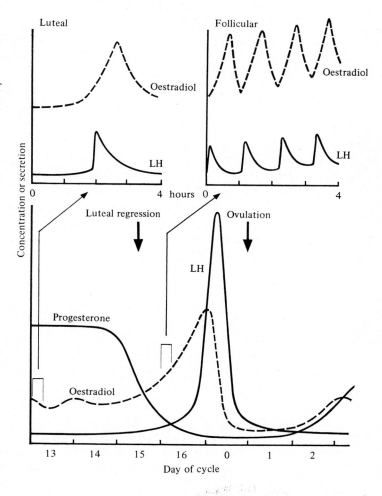

some marsupials high levels of prolactin can inhibit progesterone secretion by the corpus luteum.

Follicular growth

Folliculogenesis is a continuous process which occurs in an uninterrupted fashion until the stock of primordial follicles is exhausted (see Book 1, Chapter 2, Second Edition). Only a minority of follicles develop to a stage at which they ovulate, while the vast majority become atretic. The mechanism by which a follicle is selected for ovulation is one of the most intriguing and as yet unsolved biological mysteries. Each species has a characteristic rate of ovulation which can also be influenced by genetic, seasonal, nutritional and other environmental factors. It is likely that the hormonal requirements for follicular growth are very precise, and depend on the stage of development of the follicle. As follicular growth is asynchronous, it follows that on any one day the gonadotrophin secretion will only be optimal for a limited number of follicles. Indeed, in the human and other species in which a single ovulation is usual, there will only be a single follicle selected.

In higher primates selection of this follicle probably occurs during the first few days after the onset of menstruation (Fig. 5.17). Once selected, the follicle that will eventually ovulate becomes 'dominant', i.e. it grows

Fig. 5.17. The growth of the Graafian follicle in the follicular phase of the human menstrual cycle. (From D. T. Baird. *J. Reprod. Fert.* (in press).)

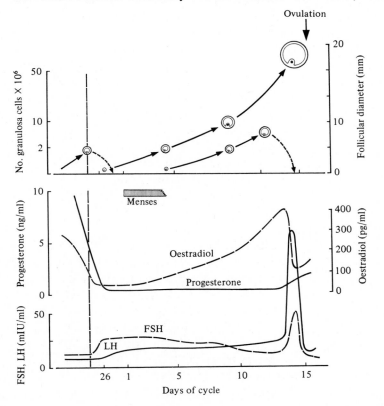

and differentiates at an exponential rate while the other antral follicles become atretic. In women it has been calculated that the follicle that will ovulate has a diameter of approximately 2 mm on Day 1 of the cycle, increasing to about 23 mm by the time of ovulation 14 days later. Associated with this growth there is a 100-fold increase in the volume of follicular fluid and the number of granulosa cells increases from 0.5×10^6 to 50×10^6. The dominant follicle secretes increasing amounts of oestradiol-17β, so that the levels in the blood rise exponentially. The rising level and increased frequency of LH discharge that occurs throughout the follicular phase of the cycle is responsible for stimulating this increase in oestrogen secretion from the dominant follicle.

The increase in growth of the pre-ovulatory follicle occurs while the concentration of FSH in the peripheral blood is falling. It seems likely that the dominant follicle becomes relatively less dependent on circulating levels of FSH since the follicular fluid itself now contains a high concentration of FSH and oestradiol. This illustrates the importance of the intra-follicular environment for growth and development of the oocyte and granulosa cells.

When the dominant follicle has reached maturity, the secretion of oestradiol is sufficient to induce a positive feedback effect and a massive discharge of pituitary LH occurs. LH induces a series of changes in the structure and function of the follicle, as well as stimulating the resumption of meiosis by the oocyte (Book 1, Chapter 2, Second Edition). The basement membrane separating the thecal and granulosa cells is disrupted, haemorrhage occurs from the ruptured thecal capillaries, and the cavity of the follicle becomes rapidly invaded with new blood vessels. The granulosa cells undergo a change known as luteinization, which involves an increase in the amount of the cytoplasm and the development of smooth endoplasmic reticulum and lipid inclusions, structures which are characteristic of steroid-secreting cells (see Chapter 7). Luteinization of granulosa cells is essential for the formation of a normal corpus luteum which in turn will provide the environment suitable for nourishment and implantation of the fertilized ovum.

Ovulation

The pre-ovulatory follicle in most species is easily recognizable by its large size and highly vascular appearance. The surge of LH induces changes in its biochemistry and structure which result in rupture of the follicle wall and extrusion of the ovum surrounded by the cumulus mass. Ovulation has been observed directly in a number of species and has been induced *in vitro* in the perfused rabbit ovary. Rupture is preceded by thinning of the follicle wall and overlying surface epithelium which produces a transparent area or stigma. Although there is a significant increase in the volume of follicular fluid, there is no increase in intra-follicular pressure prior to rupture. At the moment of ovulation the cumulus mass is extruded slowly together with a quantity of rather viscous follicular fluid (Book 1, Chapter 2, in this Edition).

There is now substantial evidence that the changes in the wall of the follicle preceding rupture are due to the release of collagenase enzymes. In the rabbit there is a large increase in the number of lysosomes in both epithelial and thecal cells underlying the stigma. As noted above, LH stimulates the production of $PGF_{2\alpha}$ and PGE_2 in pre-ovulatory follicular fluid, and ovulation can be blocked by the administration of drugs such as indomethacin which inhibit the synthesis of prostaglandins. It may be that $PGF_{2\alpha}$ stimulates the release of collagenase enzymes by the cells lining the follicle. The process is further advanced by the proteins released by this initial digestion provoking an inflammatory response, with leukocytic infiltration and release of prostaglandins and histamine. These substances inhibit repolarization of the connective tissue matrix and probably increase the local blood supply to the follicle, thereby maintaining the intra-follicular pressure. The granulosa cells are also known to produce another proteolytic enzyme, plasminogen activator, the synthesis of which is stimulated by FSH. All these changes result in the connective tissue of the follicle wall and the ground substance of the cumulus oophorus being degraded, so that follicle rupture ultimately occurs, and the oocyte is released.

These marked changes which occur in the follicle immediately prior to rupture are due to the action of LH on cells that have previously been primed with FSH and oestradiol; LH will only induce ovulation in those follicles which are sufficiently developed for their granulosa cells to have acquired receptors for LH. LH interacts with these receptors to stimulate the secretion of progesterone and initiate the series of morphological and functional changes known as luteinization. Antral follicles that are immature or in the early stages of atresia do not ovulate in response to LH but instead the process of atresia is accelerated. Atresia is poorly understood, but the stimulation of androgen production by the thecal cells, which contain LH receptors at all stages of development, is probably important.

The corpus luteum
The corpus luteum, so named by Marcello Malpighi in 1697 because of its yellow appearance in the cow, is formed from the cellular components of the follicle following ovulation. It is a distinctive structure, recognized initially by its haemorrhagic appearance, which changes depending on the species to a liver, yellow or pale cream colour within a few days. The various functional categories of corpora lutea are discussed in Chapters 6 and 7.

A normal corpus luteum will only form if the follicle has acquired an adequate number of granulosa cells with a high concentration of LH receptors. Following the pre-ovulatory LH surge, mitosis of the granulosa cells ceases so that the increase in weight of the corpus luteum following ovulation is due to an increase in the size of individual cells rather than an increase in their number. The pre-ovulatory LH surge induces a temporary loss of LH receptors from the follicular cells through a process of 'desensitization', but within a few days, stimulated by prolactin, the cells of the corpus luteum acquire a full complement of LH receptors which are essential for normal steroid secretion. Thus in most species the corpus luteum is dependent on the secretion of a luteotrophic complex of LH and prolactin. In the rabbit, however, the corpus luteum can only be maintained by oestradiol, receptors for which can easily be identified in the cytoplasm of the luteal cell. The role of oestradiol in maintaining the corpora lutea in other species, however, is still unclear.

In all species the main secretory product of the corpus luteum is progesterone – the hormone of pregnancy. However, in some primates the corpus luteum secretes a wide variety of other steroid hormones including androgens and oestradiol. Indeed, the secretion of oestradiol by the corpus luteum in women is nearly as great as that of the pre-ovulatory follicle.

In many non-primate species, luteal regression is induced by the secretion from the uterus of $PGF_{2\alpha}$, which reaches the ovary by a countercurrent mechanism involving the utero-ovarian vein and ovarian artery (see above). $PGF_{2\alpha}$ inhibits the secretion of progesterone by interfering with the ability of LH to activate the adenylate cyclase enzyme

Fig. 5.18. Proposed site of action of $PGF_{2\alpha}$ on the luteal cell. $PGF_{2\alpha}$ prevents the ability of LH to stimulate production of cAMP by blocking the interaction of the LH receptor complex with the catalytic unit. (From K. M. Henderson and K. P. McNatty. *Prostaglandins*, **9**, 779–97 (1975).)

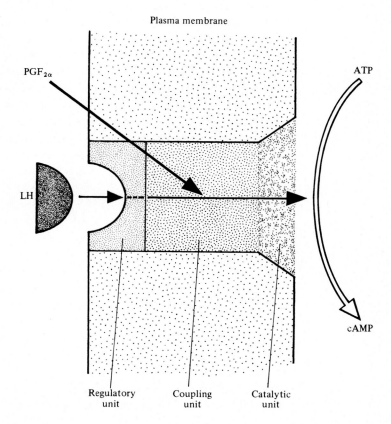

through coupling to its receptor (Fig. 5.18). The corpus luteum becomes more sensitive to the luteolytic action of $PGF_{2\alpha}$ as it ages; in many species $PGF_{2\alpha}$ cannot induce regression of the corpus luteum within a few days of its formation, probably because many LH receptors are occupied by LH. In primates, the uterus has no influence on the function of the corpus luteum, and the mechanism of luteal regression is unknown. Local injection of oestradiol directly into the primate corpus luteum causes its premature demise, and so the local production of oestradiol or prostaglandin within the corpus luteum may be responsible for luteal regression in these species.

The menopause

Although the stock of oocytes is finite and is being continually depleted by atresia and ovulation, under natural conditions most animals die long before the ovary becomes devoid of oocytes. The human female is an obvious and almost unique exception to this, and in developed countries today women may expect to live approximately 25 years after the cessation of menstrual cycles at the menopause. The post-menopausal ovary is much smaller than that of women during reproductive life since it contains no healthy follicles and no corpora lutea. The stroma contains islands of active-looking cells which are the source of the small quantities of androstenedione and testosterone secreted by the post-menopausal ovary. After the menopause neither the ovaries nor adrenals secrete significant quantities of oestrogen, but small amounts are produced in the liver and fatty tissues from androgen precursors. The absence of follicles results in a marked decline in levels of ovarian steroids (and inhibin), so the secretion of FSH and LH rises steeply (Fig. 5.19). The concentration of FSH may start to rise a few years before the last menstrual period as the number of oocytes in the ovaries is reduced, and this may account for the increased incidence of dizygotic twins which occur in older women (Fig. 5.20).

Fig. 5.19. Change in the concentration of gonadotrophins and ovarian steroids around the menopause in a woman aged 48 years. Note the fall in oestradiol concentration results in the last menses and is followed by a marked rise in the concentrations of FSH and LH. (From P. F. A. Van Look, H. Lothian, W. M. Hunter, E. A. Michie and D. T. Baird. *Clin. Endocrinol.* **7**, 13–31 (1981).)

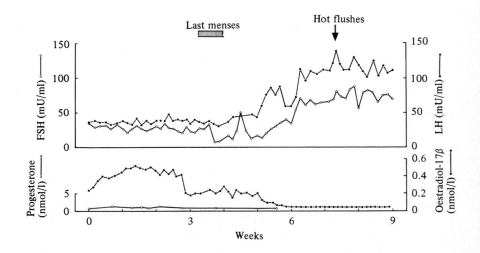

To sum up. Ovaries of all species function cyclically, alternating phases of follicular development leading to ovulation (follicular phase), followed by formation of a corpus luteum in preparation for pregnancy (luteal phase). This cyclicity is in marked contrast to the male in which spermatogenesis often continues in an uninterrupted fashion from puberty until death. In the female, although some degree of folliculogenesis continues throughout periods of reproductive quiescence, e.g. seasonal or post-partum anoestrus, the final pre-ovulatory stages of follicular development are suppressed so that sexual cycles do not occur.

The occurrence of ovarian cycles is dependent on the formation of two transitory endocrine structures – the mature Graafian follicle, which is transformed after ovulation into a corpus luteum. The length of the cycle in many species is largely dependent on the life-span of the follicle and corpus luteum, which determine the feedback effects of ovarian hormones on the hypothalamus, anterior pituitary and uterus, as well as regulating cell division throughout the reproductive tract. Thus study of the ovary must take into account its constantly changing endocrine components. Only a mature Graafian follicle secretes sufficient oestrogen to excite the hypothalamo-pituitary unit to discharge sufficient LH to cause follicular rupture and formation of the corpus luteum. For each species the length of the luteal phase tends to be fairly constant; since the secretory products of the corpus luteum have an inhibitory effect on the secretion of pituitary gonadotrophins, the corpus luteum in turn regulates the degree of

Fig. 5.20. Influence of age on incidence of twinning in Nigeria. (From P. P. S. Nylander. In: *Human Multiple Reproduction*. W. B. Saunders; London (1975).)

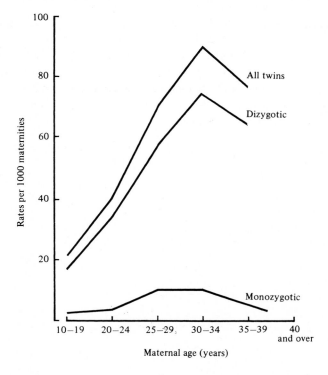

follicular development. The essential feature of the ovary is that its changing anatomical compartments result in an unstable endocrine equilibrium, which is reflected in a succession of oestrous or menstrual cycles if pregnancy does not occur.

Suggested further reading

Gonadotrophic control of follicular development and function during the oestrous cycle of the ewe. D. T. Baird and A. S. McNeilly. *Journal of Reproduction and Fertility*, Supplement **30**, 119–33 (1981).

The demonstration that $PGF_{2\alpha}$ is the uterine luteolysin in the ewe. J. R. Goding. *Journal of Reproduction and Fertility*, **38**, 261–72 (1974).

Gonadotrophins and pre-antral follicular development in women. G. T. Ross. *Fertility and Sterility*, **25**, 522–43 (1974).

Folliculogenesis in the primate ovarian cycle. G. S. di Zerega and G. D. Hodgen. *Endocrine Reviews*, **2**, 27–49 (1981).

Control of Ovulation. Ed. D. B. Crighton, N. B. Haynes, G. R. Foxcroft and G. E. Lamming. Butterworths; London (1978).

Conception in the Human Female. R. G. Edwards. Academic Press; London (1980).

The Vertebrate Ovary. Ed. R. E. Jones. Plenum Press; New York (1978).

The Ovary. Hannah Peters and K. P. McNatty. Granada; London (1980).

Reproductive Endocrinology, Ed. S. S. C. Yen and R. B. Jaffe. Saunders; Philadelphia (1978).

The Ovary, volume III, *Regulation of Oogenesis and Steroidogenesis*. Ed. Lord Zuckerman and Barbara J. Weir. Academic Press; London (1977).

6

Oestrous and menstrual cycles

R. V. SHORT

Oestrus is like baptism: it is an outward and visible sign of an inward and invisible event, in this case ovulation. It is a behavioural strategy to ensure that the female is mated at the time of ovulation. Not only does she become highly attractive to the male at this time, but she is also receptive to his advances, and may even actively seek him out – a behaviour known as proceptivity. Thus a tethered ewe 'in heat', or 'in season', will be located and mated by the ram, but equally a tethered ram will be sought out by the oestrous ewe.

The origin of the word oestrus is somewhat tortuous. It comes from the Greek word οἶστρος (*oistros*[1]), meaning a gadfly, or warble fly, a member of the Family Oestridae (Fig. 6.1) whose buzzing hum will drive herds of cattle into a frenzy during the summer months when it is active, making them toss their tails into the air, run riot and gad about. Perhaps they think

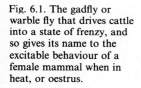

Fig. 6.1. The gadfly or warble fly that drives cattle into a state of frenzy, and so gives its name to the excitable behaviour of a female mammal when in heat, or oestrus.

---

[1] This was rendered into Latin as 'oestrus' (pronounced 'eestrus').

it is a stinging bee; but in fact the female warble fly has no sting. She lays her eggs on the hairs of the animal's legs and the eggs are ingested when the cow licks itself during grooming; the larvae ultimately migrate beneath the skin of the back, forming large abcesses or warbles that eventually burst to release the larvae onto the ground. The word 'gad', the Anglo-Saxon equivalent to the Greek *oistros*, is an old Norse word for a rod used to drive cattle. Since a cow in heat, like one pursued by a gadfly, is hyperactive, frequently bellowing and charging around the place, it was natural that the Greeks would use the word *oistros* to describe her behaviour at this time.

Menstruation gets its name from the fact that it tends to recur at monthly intervals, and the myths, legends and taboos that have grown up around it are legion. Aristotle believed that the seed of the male caused the menstrual blood or 'catamenia' to coagulate into an egg, from which the embryo developed. Milk was thought to be formed from menstrual blood, since it was observed that lactating women did not menstruate. Menstrual blood was said to be able to ward off evil and cure diseases, extinguish fires, temper metals and protect men against wounds in battle. However, the predominant sentiment in most cultures has been to regard menstrual blood as unclean, or even venomous. The Church of England to this day has a service in the prayer book for 'The Churching of Women', a ceremony to be performed after a woman had cleansed herself by having her first post-partum menstruation. Hindu women are not supposed to prepare their husband's food when they are menstruating; Moslem women cannot pray in a mosque during menstruation, and this may pose particular problems during the annual feast of Ramadan, or when on pilgrimage to Mecca. Some Buddhists also think it wrong to enter a temple during menstruation.

Western ideas about menstruation were dominated by the concept of 'repletion of the body'; a woman's body was thought to fill up with an excess of blood which had to be discharged once a month if she was to avoid serious illness. This was finally disproved by an American physician, John Davidge, who showed in 1814 that a woman might only lose about eight ounces of blood in a typical menstrual flow, whereas removal of much larger quantities of blood by venesection did not interfere with menstruation at all. He concluded that menstruation was 'attributable to a peculiar condition of the ovaries serving as a source of excitement to the vessels of the womb, rather than to the doctrine of repletion of the body'. But old ideas die hard, and many lay people still regard menstruation as a means of getting rid of something undesirable. This is reflected in the vernacular, with menstruation usually being referred to by some derogatory term, such as 'the curse'.

Oestrous and menstrual cycles

Every mammalian species shows oestrous behaviour at ovulation, with one notable exception, the human. Contrary to some earlier reports, there is no evidence to suggest that a woman's behaviour changes in any way at around the time of ovulation. She is no more attractive, receptive or proceptive to men, and intercourse is no more frequent, than at any other time in the cycle. Although a woman may experience subjective symptoms of impending ovulation, such as increased elasticity (*Spinnbarkeit*) of cervical mucus, and abdominal pain from the enlarging Graafian follicle (*Mittelschmerz*), there are no outward anatomical, histological or behavioural changes that accompany ovulation. Unlike all other mammals, a woman is potentially attractive to the male, and may be receptive to his advances, at any time from puberty to old age, although any man who assumes that women are constantly in oestrus is in for a surprise. The fact that a woman's sexual behaviour is no longer a slave to her hormones must have been of enormous significance in human evolution. Not only has human sexuality become a cohesive force in the maintenance of male–female pair bonds, but the lack of overt oestrus has made communal living a less stressful lifestyle than might otherwise have been the case (see Book 8, Chapters 1 and 2, First Edition).

In the absence of oestrus, we clearly cannot talk about a human oestrous cycle. The only outward event that reveals the inherent cyclicity of a woman is menstruation. This is a physiological as opposed to a behavioural phenomenon (although it may have behavioural manifestations like pre-menstrual tension) that is restricted to humans and Old World (African and Asian) primates, and the old adage that menstruation represents 'the womb weeping for its lost lover' is particularly apt.

Our first understanding of the hormonal basis for menstruation came from studies carried out in the mid 1930s on ovariectomized rhesus monkeys given hormone replacement therapy. It was shown that treatment with either oestrogen or progesterone followed by hormone withdrawal would produce menstruation. When oestrogen and progesterone were given in combination, the withdrawal of oestrogen alone was without effect, whereas the withdrawal of progesterone alone would produce menstruation even though the oestrogen was maintained. Thus in a normal cycle, menstruation is caused by the declining secretion of oestrogen and progesterone by the regressing corpus luteum; in women on the combined oestrogen plus progestogen contraceptive pill, menstruation follows the cessation of pill taking; in women who fail to ovulate, menstruation may occur following a decline in oestrogen secretion from the Graafian follicle as it becomes atretic.

These hormonal changes are thought to produce changes in the endometrial vasculature, although the precise biochemical mechanism remains to be determined. In 1940, Markee hit upon the ingenious idea of transplanting pieces of endometrium from female rhesus monkeys into the anterior chamber of their eyes, thus making it possible to observe the process of menstruation directly. He concluded that menstruation was always preceded by vasoconstriction of the spiral arterioles at the base of the endometrium; this resulted in endometrial necrosis and damage to the endothelium of the endometrial blood vessels, so that when the spiral arterioles relaxed once more, haemorrhage occurred. We now know that this haemorrhage is facilitated by the local release of vasodilators such as histamine, bradykinin and prostacyclin and other prostaglandins. Not only are prostaglandins present in the menstrual fluid, but the administration of prostaglandin synthetase inhibitors will reduce the volume of blood lost at menstruation (see Book 7, Chapter 3). It is thought that the reason why menstruation is almost entirely confined to the Old World primates is that they are the only ones that have spiral arterioles in the endometrium. Disorders of menstruation are among the commonest gynaecological complaints, and 'spotting' or 'breakthrough bleeding' is one of the major drawbacks of many steroidal contraceptives, particularly those pills, implants and injections containing only progestogens. If we knew more about the underlying physiology of menstruation, we might be able to adopt a somewhat less empirical approach to the solution of these problems.

We often do not appreciate just how late in the day it was before we knew when in the menstrual cycle a woman ovulated. In 1842 the great French anatomist, Professor Pouchet of Rouen, firmly stated in his Ten Fundamental Laws of Reproduction that women ovulated at menstruation, and he advocated a 'rhythm' method of contraception based on this fact. But it was another Frenchman, Professor Raciborski in Paris, who pointed out in the following year that women could not possibly ovulate at menstruation; he had observed that girls whose wedding took place immediately after a menstrual period would often conceive without menstruating again, whereas girls marrying late in their cycles almost invariably menstruated again before they conceived. It was a Japanese gynaecologist, Ogino, who in 1930 provided the first conclusive proof that ovulation occurred midway between the two menstrual periods by laparotomizing women at various stages in the cycle. This was soon confirmed by Professor Knaus in Austria, and both of them advocated a 'safe period' regime for contraception.

The sharp distinction between the human menstrual cycle and the oestrous cycle of animals becomes blurred when we try to find an appropriate terminology to describe the cycles of Old World primates, such as the chimpanzee or rhesus monkey. The laboratory worker, studying these animals at close quarters in cages, is impressed by the fact that, like women, they menstruate, and so talks in terms of their menstrual cycles. The field worker, on the other hand, viewing the animals at a distance through binoculars, cannot detect menstruation. Instead, he observes that the animals come into behavioural oestrus, and since this is often accompanied by major anatomical changes, such as perineal tumescence and reddening of the sexual skin, it is particularly easy to detect. He can therefore only describe these events in terms of an oestrous cycle, even though they are taking place in a menstruating species.

A further complication is that oestrus or menstruation can occur in the absence of ovulation. Anovular cycles are particularly common in women, chimpanzees and rhesus monkeys following the first menstruation, or menarche. The reason for this ovulatory failure appears to be that the

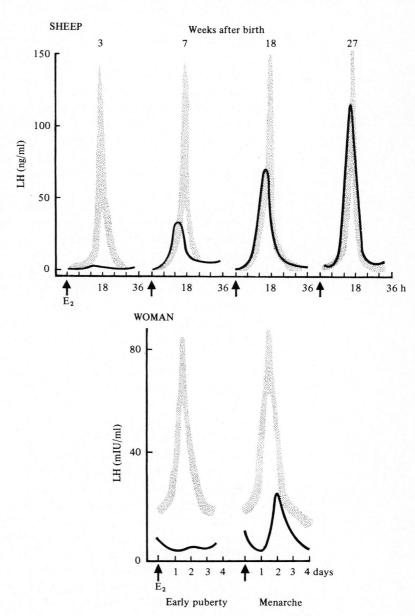

Fig. 6.2. Development of the positive feedback mechanism with age in sheep and women. An oestrogen injection (E_2) causes the discharge of increasing amounts of luteinizing hormone from the pituitary gland. Stippling represents the adult response. This mechanism matures before puberty in sheep, but after puberty (menarche) in humans. (From M. C. Levasseur and C. Thibault. *De la Puberté à la Sénescence*. Masson; Paris (1980).)

so-called positive feedback mechanism, by which rising ovarian oestrogen secretion normally triggers an ovulatory discharge of luteinizing hormone from the pituitary, does not mature for a year or two after puberty in these primates (Fig. 6.2). Thus follicles develop in the ovaries, secrete oestrogen, and stimulate development of the endometrium. In the absence of a positive feedback 'circuit breaker' to induce ovulation, the follicles eventually become atretic and stop secreting oestrogen. This oestrogen withdrawal eventually induces menstruation, although the cycles tend to be longer than normal, and rather variable (Fig. 6.3).

Fig. 6.3. Length of the menstrual cycle in women from first menstruation (menarche) at about the age of 13 to last menstruation (menopause) at about the age of 50. The solid line shows median cycle lengths, and 90% of cycle lengths fall within the upper and lower broken lines. Note the high incidence of extended cycles soon after menarche, and before the menopause. (From A. E. Treloar *et al. Internat. J. Fertil.* **12**, 77–126 (1967).)

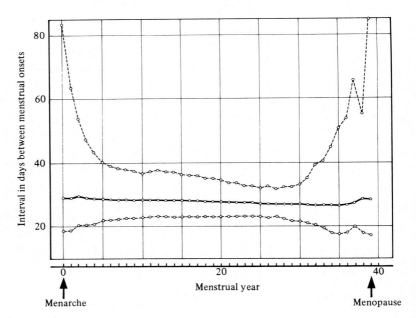

Fig. 6.4. Urinary oestrogen and pregnanediol excretion in a girl going through puberty between the ages of 13 and 15. Note that waves of oestrogen excretion preceed menarche, and that the first ovulation, indicated by a pregnanediol excretion of 2mg/24h, occurs about a year after menarche. (Adapted from J. B. Brown *et al. J. Biosoc. Sci., Suppl.* **5**, 43–62 (1978).)

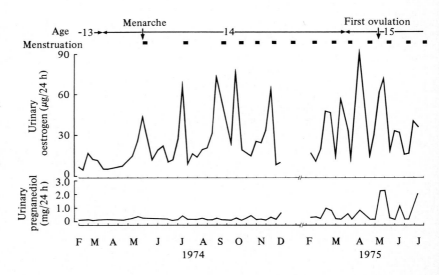

These post-pubertal anovular cycles may have been of some adaptive significance in primitive man and higher primates, since they allowed the female to select an appropriate mate by sexual trial-and-error, without paying the genetic price of conceiving to the first male that came along. Fig. 6.4 shows how cyclical ovarian oestrogen secretion in a girl starts before the first menstrual period, and continues for a year after menarche before the first ovulation takes place. Fig. 6.5 describes a similar sequence of events in a chimpanzee living in the wild in the Gombe Stream Reserve in Tanzania; here the degree of perineal tumescence serves as a bioassay for oestrogen secretion, and pregnancy is proof of ovulation. Although copulations occurred with many different males at peak tumescence when the chimpanzee was in oestrus, almost three years elapsed between menarche and conception, presumably because of the anovular cycles.

Another complication in the interpretation of cycles is that ovulation can sometimes occur in the absence of oestrous behaviour, a condition described by that delightful mixed metaphor, 'silent heat' (in reality, a 'silent ovulation'). One of the best examples of this is seen in sheep, which are seasonal breeders. The first ovulation of the mating season in the autumn is usually not accompanied by oestrus, because the ewe requires progesterone-priming before oestrogen can induce full oestrous behaviour

Fig. 6.5. Cycles of perineal tumescence in a chimpanzee living in the wild, from menarche at age 11 to first pregnancy. Note that irregular cycles of tumescence precede the first observed menstrual bleed, and that cycles continue, with matings when in oestrus at peak tumescence, for almost 3 years before conception. Some cycles also continue on into pregnancy. Black bars indicate menstruation. Data supplied by Dr Caroline Tutin.

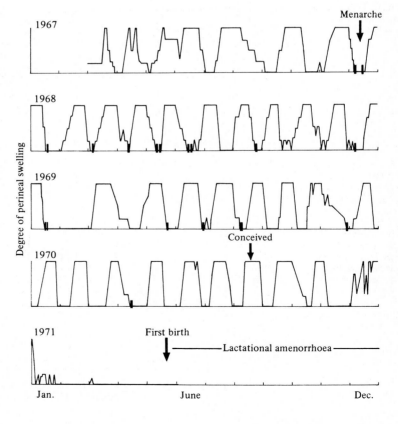

(Fig. 6.6). Perhaps these silent heats may nevertheless serve to alert the rams to the fact that the mating season is about to begin, and hence heighten the intensity of their pre-copulatory rutting behaviour.

In contrast to silent oestrus, the only occasion when menstruation fails to occur following an ovulation is when the individual has become pregnant. However, it is interesting to note that oestrous-like perineal swellings continue to occur during the early months of pregnancy in chimpanzees (see Fig. 6.5). Although the presence of the embryo prevents luteal regression and hence menstruation, presumably cycles of follicular development and oestrogen secretion continue to occur. These follicles are destined to become atretic, since progesterone inhibits the positive feedback mechanism and hence prevents ovulation.

Are cycles a normal feature of reproduction?

It has become a time-honoured practice to regard the cycle as the 'normal' form of reproductive activity, and pregnancy as something of a special event. Thus we talk about the normal menstrual cycle, or the normal oestrous cycle, as if this was nature's first priority in reproductive design. Of course, nothing could be further from the truth. *Cycles are only a consequence of infertility*; they are nature's way of making another bid for pregnancy. Anybody who has attempted to study wild animals, large or small, will know that oestrous cycles hardly ever occur (other than in the special case of adolescent sterility in primates, which we have just mentioned). Most females conceive at the first ovulation, and hence never have the need, or the opportunity, to exhibit repeated cyclical behaviour.

If the repeated menstrual or oestrous cycles of women and domestic and laboratory animals are a result of human intervention, and hence a relatively recent phenomenon, they will not have been subjected to the refining process of natural selection. Could too-often repeated cycles actually be harmful? Are there hazards in nulliparity? There is some evidence, particularly in women, to suggest that this may in fact be the case.

Fig. 6.6. The oestrous cycle of the ewe, showing how the first ovulation of the mating season in the autumn is not accompanied by behavioural oestrus since progesterone-priming is necessary before oestrogen can produce its behavioural effects.

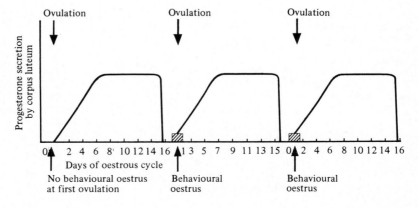

Our early ancestors, like present-day hunter–gatherers, would have had one or two years of menstrual cycles during the period of adolescent sterility from menarche to first conception, and then nine months of pregnancy amenorrhoea, immediately followed by 2–3 years of lactational amenorrhoea. Then there might be as many as two or three more menstrual cycles before the next conception, several more years of lactational amenorrhoea, and so on. Thus a fertile woman would give birth to about five children, but would only experience 30 or 40 menstrual cycles during her entire reproductive life-span. Compare this to the situation in present-day contracepting Western woman, with menarche at 13 and the menopause at 50, and maybe two pregnancies, each followed by only three months of amenorrhoea because of the early abandonment of breast-feeding. She would experience over 400 menstrual cycles (see Fig. 6.7). Since each menstrual cycle produces changes in the ovaries, the uterus and the breasts, it is here that we must look for evidence of abnormalities.

The evidence that repeated menstrual cycles may be harmful is least conclusive in the case of the ovaries, although we do know that nulliparous women, who must have had a continuous succession of cycles, have a greater risk of ovarian cancer than parous women. As far as the uterus is concerned, the more menstrual cycles a woman has, the more likely she is to get endometriosis, a condition in which small pieces of the endometrium shed at menstruation pass up the Fallopian tubes to enter the peritoneal cavity. Here they become attached to any of the pelvic structures, and may cause painful symptoms when the tissue undergoes cyclical bleeding in synchrony with the shedding of the normal endometrium at menstruation. There is also some evidence to suggest that nulliparous women are more likely to have uterine fibroids, and develop carcinoma of the endometrium.

Fig. 6.7. The incidence of menstrual cycles during the reproductive life of a typical hunter–gatherer woman from the Kalahari, who begins intercourse at puberty and has long periods of lactational amenorrhoea separating successive births, compared to a Western woman who has abandoned breast-feeding and uses modern contraceptives. Numbered squares indicate pregnancies. The earlier age of menarche in Western women is probably due to improved nutrition.

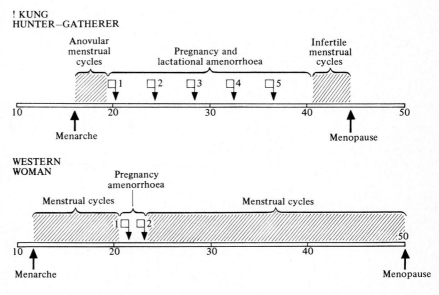

There is a considerable morbidity associated with the process of menstruation itself. Six per cent of women will lose more than 100 ml of blood during a period (see Fig. 6.8), and although this may be of little consequence for a healthy well-nourished Western woman, it may be a serious matter for an undernourished, protein-deficient woman in the Third World, having to perform arduous physical work and cope with concurrent hookworm and malarial infestations that further deplete her haemoglobin stores. Many women also suffer significant psychological and physical distress at the time of menstruation, with pre-menstrual tension, backache and abdominal pain that may be severe enough to require bed rest or days off work. We should also remember that the incidence of hysterectomy reaches staggering proportions in some communities, one of the principal indications for the operation being painful, irregular or uncontrollable menstruation. In the United States, about 37 per cent of women born at the turn of this century had had an hysterectomy before the age of 70; in England and Wales the incidence is about 6 per cent, although if present trends continue it will rise to about 19 per cent very soon.

The clearest example of a hazard of nulliparity relates to breast cancer, which affects at least one woman in twenty in developed countries, and is by far the commonest cancer of women. Oestrogen appears to be the culprit, and the breast is exposed to unopposed oestrogenic stimulation in the first half of every menstrual cycle. Careful epidemiological studies have shown that the risk of getting breast cancer is increased by an early age of menarche, and a late age at menopause, both factors that effectively increase the number of menstrual cycles to which a woman is exposed. The first pregnancy appears to have a protective effect, and the earlier it occurs,

Fig. 6.8. Volume of blood lost at menstruation from a randomly selected sample of 476 Swedish women aged 15–50. (From L. Hallberg *et al. Acta obstet. gynec. scand.* **45**, 320–51 (1966).)

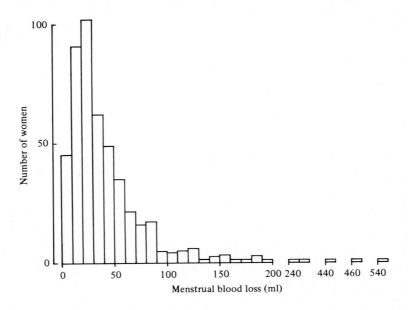

the greater the protection (Fig. 6.9), although it does not seem to matter how many subsequent pregnancies there are. Breast cancer is therefore particularly common in nuns. The incidence of breast cancer is significantly lower in developing countries (Fig. 6.10), and much of this can be explained by the differences in reproductive life history. Recent epidemiological studies have shown that the probability of developing breast cancer is highly correlated with the number of menstrual cycles a woman has experienced in her life.

Unfortunately we lack studies in animals on the effects of long periods of enforced infertility on morbidity and mortality rates, and on subsequent fertility; however it does seem that repeated menstrual cycles represent an unhealthy form of infertility for women. In designing contraceptives for the future, our goal should surely be to achieve a more healthy state of infertility. We must ask ourselves 'What is the optimum endocrine environment for a breast that we do not want to lactate, an ovary that we do not want to ovulate, and a uterus that we do not want to menstruate?' We already have contraceptives, like the long-acting gestagen 'Depo-Provera', which will produce amenorrhoea as a by-product of their contraceptive effect; others that also produce amenorrhoea, like the

Fig. 6.9. Protective effect of the first pregnancy on subsequent incidence of breast cancer. Those who first give birth before the age of 30 have a lower risk of getting breast cancer in later life than nulliparous women, whereas first births after 30 increase the risk. (From B. MacMahon *et al. J. Natn. Cancer Inst.* **50**, 21–42 (1973).)

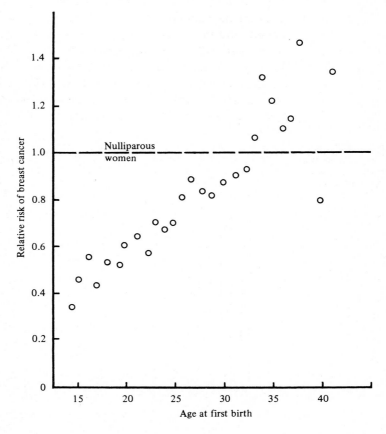

releasing-hormone agonists and antagonists, are under development. Women in developed countries should seriously ask themselves the question 'Why menstruate?' If the conventional oestrogen/gestagen oral contraceptive pill is taken for an extended period of time, and not just for 21 days, menstruation can be suppressed, and our own studies in Edinburgh showed that this was highly acceptable to a large number of middle-class women. Even in developing countries, many women might welcome a longer period of time following childbirth before their menstrual periods resumed, although they might regard amenorrhoea prior to their first pregnancy as a potential threat to their future fertility.

Classification of cycles

It was Walter Heape, a Cambridge zoologist, who first attempted in 1900 a systematic classification of the many different types of oestrous cycles seen in mammals. He talked about pro-oestrus, the phase of follicular development preceeding oestrus; oestrus itself; metoestrus, which immediately follows oestrus; dioestrus, or the phase of luteal activity, and anoestrus. Although this terminology is still used to describe the cyclical events in laboratory rodents, it has not proved particularly useful for the larger domestic animals or the primates.

Another simple classification is to divide animals into spontaneous and induced ovulators. Spontaneous ovulators do not require the act of coitus to induce follicular rupture, and most mammals fall into this category. Induced ovulators include an odd assortment of species like the rabbit, ferret, mink, field vole, cat and camel in which coitus provokes a reflex discharge of GnRH and hence of LH from the pituitary, thus inducing

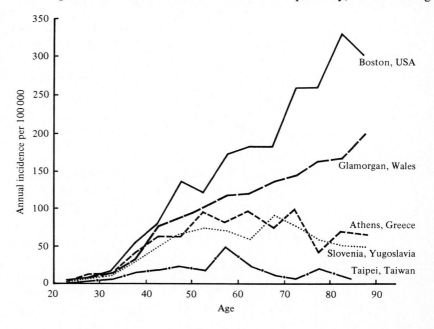

Fig. 6.10. Annual incidence of breast cancer by maternal age in developed and developing countries. (From B. MacMahon *et al.* *J. Natn. Cancer Inst.* **50**, 21–42 (1973).)

ovulation. The laboratory rat, mouse and hamster are something of a half-way house between these two extremes, since although they are spontaneous ovulators, they nevertheless require the act of coitus to produce a fully functional corpus luteum. In the absence of coitus, the corpus luteum of dioestrus only secretes for two or three days before regressing. Mating appears to stimulate the release of prolactin from the anterior pituitary, and this is necessary for full luteal activity. There may be considerable overlap between spontaneous and induced ovulators, since the time of ovulation can be advanced by copulation in some spontaneous ovulators, and it has even been suggested that this could occur in women.

Another major difference between species relates to the life-span of the corpus luteum. There are those species like the dog and ferret, and the kangaroos and wallabies, in which the life-span of the corpus luteum of the non-pregnant animal is almost identical to that of the corpus luteum of pregnancy. But in most species, including primates, ungulates, and rodents, the corpus luteum of the non-pregnant animal lasts for a much shorter time than the corpus luteum of pregnancy, the placenta having developed a luteotrophic capacity (see Chapter 7). The situation is reversed in many of the smaller marsupials like the Virginia opossum of North America and the brush possum of Australia, where the corpus luteum of the cycle lasts for longer than the corpus luteum of pregnancy (see Table 6.1).

It is at this point that difficulties in terminology begin to arise. Those working with laboratory rodents are happy with the clear distinction between three classes of corpora lutea, based on their mode of formation, life-span, and secretory activity. There are the short-lived, inactive corpora lutea of dioestrus, or corpora lutea of the cycle, formed following an oestrus at which mating has not taken place; there are the longer-lived and more functional corpora lutea of pseudopregnancy, formed as a result of a sterile mating; and there are the most long-lived and functional corpora lutea of all, namely those of pregnancy. But in most other mammals there are only two classes of corpora lutea, and rather than calling the corpus luteum in a non-pregnant wallaby, cow, sheep or woman a corpus luteum of dioestrus, or of pseudopregnancy, convention has it that we usually refer to it as the corpus luteum of the cycle, or even the non-pregnant corpus luteum.

Another type of classification of cycles that is frequently used is related to the number of cycles that occur in a year. Thus the fox or the roe deer, which only come into oestrus once each year at a predictable time, are described as being seasonally monoestrous. Unmated tammar wallabies, horses, sheep, goats, red deer and rhesus monkeys, on the other hand, which show a succession of oestrous cycles at a certain time of year, are described as seasonally polyoestrous. Those species that can show cyclical activity throughout the whole year if not mated, like rats and mice, chimpanzees and orang-utans, and cows and pigs, are described as

Table 6.1. *Types of oestrous cycles, cycle lengths (days), and duration of gestation (days) in a variety of marsupial and eutherian mammals*

Species	Type of cycle	Spontaneous (S) or Induced (I) ovulator	Non-pregnant cycle		Pregnancy
			Dioestrus	Pseudo-pregnancy	
Mouse	Polyoestrous	S	4	12	19
Rat	Polyoestrous	S	4	14	22
Golden hamster	Seasonally polyoestrous	S	4	10	16
Guinea pig	Polyoestrous	S	16		68
Rabbit	Polyoestrous	I		16	28
Ferret	Seasonally polyoestrous	I		42	42
Cat	Seasonally polyoestrous	I		30	63
Dog	Monoestrous	S	61		61
Horse	Seasonally polyoestrous	S	21		350
Cow	Polyoestrous	S	21		282
Sheep	Seasonally polyoestrous	S	16		148
Goat	Seasonally polyoestrous	S	16		148
Pig	Polyoestrous	S	21		113
Red deer	Seasonally polyoestrous	S	18		234
Roe deer	Seasonally monoestrous	S	150?		300*
Tammar wallaby	Seasonally polyoestrous	S	29		29(365)*
Red kangaroo	Polyoestrous	S	33		33(235)*
Virginia opossum	Seasonally polyoestrous	S	28		13
Brush possum	Polyoestrous	S	26		17
Camel	Seasonally polyoestrous	I	?		390
Indian elephant	Polyoestrous	S	126		624
Marmoset monkey	Polyoestrous	S	18		148
Rhesus monkey	Seasonally polyoestrous	S	28		165
Orang-utan	Polyoestrous	S	30		260
Chimpanzee	Polyoestrous	S	36		228
Gorilla	Polyoestrous	S	28		255
Human	Polymenstrual	S	28		267

* Includes embryonic diapause.

polyoestrous. Presumably we should describe ourselves as polymenstrual, although the term is seldom used. Species in which mating is confined to certain periods of the year are called seasonal breeders. However use of the phrase 'breeding season' may cause confusion, since this can refer either to the season of mating, or to the season of birth.

Animals may also differ from one another in the time of day at which ovulation occurs. It has long been recognized that rats and hamsters produce a pre-ovulatory discharge of LH only on the afternoon of the day of pro-oestrus, resulting in ovulation 12 hours later in the early hours of the day of oestrus. This fits with the nocturnal activity patterns of these animals. It has always been assumed that larger animals showed no such circadian rhythm in the timing of ovulation, but recently Bob Edwards in Cambridge, during the course of his human *in vitro* fertilization research, has established that the pre-ovulatory LH surge in women almost invariably begins in the morning (Fig. 6.11). Since the LH surge precedes ovulation by about 30 hours, this suggests that women usually ovulate in the late afternoon. If human ovulation is under circadian control, this may explain why time-zone changes in the pre-ovulatory phase of the cycle can postpone ovulation and hence extend the menstrual cycle – a common complaint amongst airline hostesses.

What biological sense can we begin to make out of this baffling array of species diversity in oestrous and menstrual cycles? For the small mammals, with large litter sizes, and short life-spans allowing them only a very limited time in which to reproduce, it is important to be able to put in a new bid for pregnancy as soon as possible if the preceding one failed. These mammals thus tend to have short cycles. The larger the animal, the longer the life-span, and hence the pressures for instantaneous reproductive success are reduced. Litter sizes are in general smaller, cycle lengths are

Fig. 6.11. Diurnal rhythm in the time of onset of the LH surge in women, as measured in urine. (From R. G. Edwards. *Nature*, **293**, 253–6 (1981).)

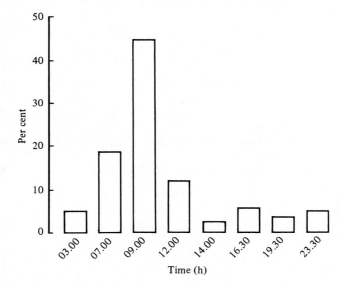

longer, and there is a longer interval between successive births. For species with really long life-spans, like man himself, it has even been necessary to go to the other extreme and put the brakes on fertility: this has been achieved by postponing the age of puberty, increasing the interbirth intervals by long periods of lactational amenorrhoea, prematurely terminating reproductive activity at the menopause, and having a rather low probability of conception (about 25 per cent) even when ovulation does occur.

Cycle lengths and signs of oestrus

In order to record the length of the oestrous cycle, it is first necessary to isolate the females from contact with a fertile male. It may then be possible to detect the animal in heat by the homosexual behaviour of the other females in the group towards her (Fig. 6.12). The dairy farmer, for example, will detect a cow that is 'bulling' by the fact that she will stand still and allow herself to be ridden by other cows in the herd. Failure of the farmer to detect oestrus is still the major cause of infertility in dairy herds that rely on artificial insemination. It has recently been shown that dogs can be trained to sniff out cows in heat; there must therefore be some pheromone present in the vaginal mucus, but its chemical composition has yet to be determined.

The horse-breeder cannot use homosexual behaviour to detect oestrus, since mares will not mount one another; he has to use a 'teaser' stallion

Fig. 6.12. Red deer hind mounting another hind in oestrus.

who is introduced to the mare over the top of a gate. If she is in oestrus, she will raise her tail, urinate, and 'wink' – a rhythmic contraction of the vulva and eversion of the clitoris. The sheep-breeder will often use a vasectomized ram whose chest has been painted with coloured grease, so that the ram leaves his mark on the rump of any ewe that has allowed him to serve her. The pig-breeder may make use of a spray can of boar pheromones, the Δ^{16}-steroids androstenol and androstenone (Fig. 6.13a), which the boar normally secretes in his saliva and sprays around the pen as he froths at the mouth and shakes his head in a frenzy of sexual excitement. When the sow smells these steroids, she becomes immobile if she is in oestrus, and will allow the farmer to press on her back, or even

Fig. 6.13. (a) Boar pheromones, the steroids Δ^{16}-androstenol and Δ^{16}-androstenone. (b) Farmer checking heat in a sow exposed to the smell of boar pheromones.

(a)   BOAR PHEROMONES

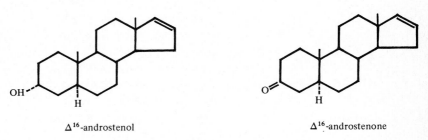

Δ^{16}-androstenol

Δ^{16}-androstenone

(b)

to ride on her (Fig. 6.13*b*). It is of interest that high concentrations of androstenol have recently been isolated from truffles, which now explains why sows were traditionally used to hunt for these underground fungi, the gourmet's delight, in the beech woods of France. These same Δ^{16}-steroids are present in a man's sweat and urine; few men can smell them, whereas a high proportion of women find the smell unpleasant, and therefore show a reluctance to cook pork from uncastrated male pigs.

One of the most enigmatic and occult oestrous cycles is that of the elephant. Even experienced elephant keepers in zoos and circuses have been completely unable to determine when a cow elephant comes into heat, and it was not until the staff at Portland Zoo in Oregon started to investigate this subject that we learned the truth. The bull Asiatic elephant checks his cows for heat by placing his trunk in the vicinity of the vulva and taking a sniff. If he detects the oestrous pheromone, whatever it may be, he then places the tip of his trunk in his mouth and exhales onto the two openings of the vomeronasal ducts, on the roof of the mouth beneath the hard palate. By studying the behaviour of a bull elephant towards unmated cows in a daily testing session, carried out for months on end, it was possible to show that oestrus lasted for about a week, and occurred at 18 week intervals, coinciding with a decline in the minute amounts of progesterone present in the peripheral blood.

The vomeronasal organ seems to be used specifically by males for savouring sexual smells, and the associated behaviour shown by the male is known as 'flehmen'. A male herbivore, when smelling a female (or her urine) in oestrus raises his head and curls the upper lip back; this in some way allows odours to reach the vomeronasal organ. This behaviour was described in picturesque language by Turberville, writing in 1576 about the sexual behaviour of red deer stags: 'For when they smell the Hynde, they rayse their nose up into the ayre, and looke aloft, as though they gave thanks to Nature which gave them so great delight'.

The dog-breeder has perhaps the easiest job of all in detecting oestrus, since the bitch gives at least 10 days warning of an impending ovulation by the oestrogen-induced tumescence of her vulva, which is accompanied by the appearance of fresh blood. This pro-oestrous bleeding is not to be confused with menstruation, and is due to an extravasation of red blood cells from the oestrogenized vagina. After about a week of these pro-oestrual changes, the vaginal epithelium becomes cornified, and the bitch comes into full oestrus for several days. She becomes disobedient to her owner, and highly attractive to all the male dogs in the neighbourhood as a result of the excretion of a pheromone, methyl-*p*-hydroxybenzoate, from her vagina. This is also present in her urine, and her urination frequency increases at the time of oestrus. Most domestic breeds of dog come into oestrus twice a year, but the timing of oestrus is very variable and unrelated to the seasons. Thus it is impossible for the dog-breeder to predict with any accuracy when the next oestrus will occur.

Perhaps the most interesting oestrous cycles are those of the primates. In some species, like baboons, macaques, and the chimpanzee, there is a marked hyperaemia and swelling of the perineal skin at oestrus, providing a flagrant visual advertisement of the animal's reproductive state, often to the embarrassment of parents taking their small children round zoos. However, closely related species, like the gibbons, the orang-utan and the gorilla, show little or no change in the vulva, and Charles Darwin himself was fascinated by this subject, admitting that 'no case interested and perplexed me so much as the brightly coloured hinder ends and adjoining parts of certain monkeys'. But Tim Clutton-Brock in Cambridge has pointed out that the explanation is disarmingly simple; swellings are only seen in terrestrial primates, and in those with polygynous mating systems. In arboreal species, long-range visual cues would be of no use because visibility is reduced to a few feet by the foliage. And in monogamous species, where the male and female are constantly in one another's company, there is likewise no need to advertise the female's condition to other males. This makes the point that we cannot begin to understand the diverse types of oestrous behaviour until we have studied the ecology of the species concerned in its natural habitat in the wild.

Historically speaking, the most intriguing oestrous cycle is that of the camel, an induced ovulator. Aristotle, writing in about 350 BC, records how female camels that were wanted for war purposes were 'mutilated', by which he meant spayed or ovariectomized. This was a common agricultural practice at that time amongst pig-farmers, who had observed that removal of the ovaries from sows prevented oestrus and pregnancy, thus allowing animals to fatten better. However, it is something of a mystery as to how the farmers empirically developed the operation in the first place, since it was 2000 years before scientists like Stensen and De Graaf in the late seventeenth century first realized that the ovaries, or 'female testicles' as they called them, had anything to do with reproduction. The clue may lie in the camel. When female camels were used in warfare, it was desirable that they should not be pregnant. However, if an induced ovulator like a camel (or a cat or a rabbit) is prevented from being mated, the result is that she is almost constantly in oestrus, as waves of follicles develop and regress with no intervening luteal phase. Unfortunately, one of the signs of oestrus in the camel is that she sits down; copulation is a leisurely and lengthy affair, often lasting many hours (Fig. 6.14). Thus if the enemy were to release a male camel amongst the non-pregnant females of the opposing army, the battle would be lost before it began. Ovariectomy in the camel may have been both man's first endocrine experiment, and his first attempt at biological warfare.

And so we come to the laboratory rodents: the rat, mouse, guinea pig and hamster. These have truly been the work-horses of the reproductive biologist in the twentieth century, and one of the reasons for their success is that they all show marked and predictable histological changes in their vaginal smears during the oestrous cycle, with a high proportion of cornified cells present only on the day of oestrus. This fact, first discovered by C. R. Stockard and G. N. Papanicolaou in the guinea pig in 1917, has provided scientists with a simple way of monitoring ovarian activity without the need for hormone assays. It also provided an invaluable bioassay for oestrogens that was essential for their isolation and chemical characterization. Rats and hamsters also show predictable behavioural changes during the oestrous cycle. If an exercise wheel is placed in the female's cage, she will show intense running activity on the day of pro-oestrus. This appears to be a reflection of her proceptive behaviour, since in an open field situation she will extend her home range enormously during pro-oestrus as she searches for a mate.

As far as women are concerned, it will be apparent from what has already been said that in the absence of any behavioural signs of oestrus, it is particularly difficult to detect when in her menstrual cycle a woman ovulates. There are two do-it-yourself methods that give some indication, though neither is particularly precise – or infallible. Measurement of the basal body temperature, taken on first waking in the morning, will usually show a rise of 0.5–1 °F after ovulation has occurred (Fig. 6.15). This is because the progesterone metabolite, pregnanediol, has a thermogenic effect, apparently altering the setting of the thermoregulatory centre in the brain. Not all women show this shift in basal body temperature, and unfortunately it cannot predict when ovulation is going to occur – it can only indicate when it has occurred. The second method depends on subjective assessment of the viscosity of cervical mucus. This is a good oestrogen bioassay, and will indicate when oestrogen-secreting follicles are present in the ovaries, although not when ovulation has occurred. For a woman who is experienced in the technique and is sufficiently dedicated to use it, it can provide a valuable indication of when ovulation is likely

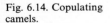

Fig. 6.14. Copulating camels.

to occur (Fig. 6.15); it forms the basis of the Billings method of 'natural' family planning.

Of course, clinicians now have available a whole range of more sophisticated methods for detecting ovulation. Measurement of urinary pregnanediol and oestrogen excretion was the traditional approach (Fig. 6.15), but in recent years, with the development of artificial insemination and *in vitro* fertilization, there has been a need for a more precise predictive indicator. Immunological measurement of LH in blood or urine is perhaps the most reliable method (Fig. 6.15), and to this can now be added visualization of the follicle and the newly formed corpus luteum by ultrasound. A useful method for establishing when ovulation has occurred is to measure progesterone in saliva; spit samples are far easier for the patient to collect than blood or urine, and since salivary concentrations of progesterone (and oestrogen, cortisol and testosterone) closely mirror the circulating concentrations of free hormone, salivary assays may be used increasingly for diagnostic procedures in the future.

Before leaving the subject of cycles, one obvious question springs to mind: what determines cycle length? Is it some inherent rhythm in the hypothalamus, or does the rhythm generator lie in the ovary itself? Since the ovaries are dependent on pituitary gonadotrophin secretion, and ovarian steroid feedback in turn modifies hypothalamic and pituitary activity, it is difficult to determine in such a circular system where cyclicity originates.

Certainly the time-lags in the ovarian response to pituitary gonadotrophic stimulation are very important in some species in determining cycle length. The life-span of the corpus luteum is one of the principal determinants,

Fig. 6.15. Changes in basal body temperature, blood FSH and LH concentrations, viscosity of cervical mucus, and total urinary oestrogen and urinary pregnanediol excretion in a woman during the course of 6 successive ovulatory menstrual cycles. A mucus score of 5 or more, indicating stretchy, wet and slippery mucus, indicates potential fertility and correlates well with maximal follicular oestrogen secretion and the pre-ovulatory LH surge. Black bars indicate menstruation. (From J. B. Brown *et al.* Ovulation Method Research and Reference Centre of Australia (1981).)

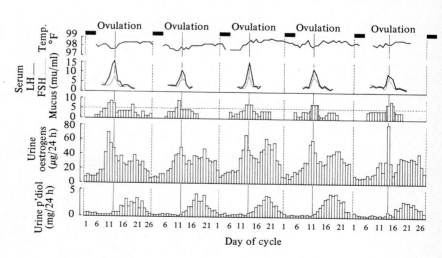

since progesterone blocks the positive feedback effects of oestrogen, and hence inhibits the LH surge and ovulation. Another ovarian factor is the time taken for a follicle to mature to the point of ovulation after the corpus luteum has regressed; in sheep it takes 2–3 days from luteal regression or surgical removal of the corpus luteum to the next ovulation, so that the 16-day oestrous cycle is composed of a luteal phase of 4 days and a brief 2-day follicular phase. In women and rhesus monkeys, on the other hand, it normally takes 12–14 days from luteal regression or corpus luteum enucleation to ovulation, so the 28-day menstrual cycle is composed of a 12- to 14-day phase of follicular maturation and a 14-day luteal phase.

A dramatic example of the ability of the ovary to determine cycle lengths in the absence of any hypothalamic rhythmicity comes from the experiments of Ernst Knobil and his colleagues on the rhesus monkey. They surgically destroyed the arcuate–ventromedial region of the hypothalamus that is responsible for GnRH secretion, hence depriving the pituitary gonadotrophs of all tropic stimulation. Gonadotrophin levels fell to zero, and all ovarian activity ceased. The animals were then pulsed once an hour for weeks on end with exogenous GnRH. Follicles developed in the ovaries, the rising levels of oestrogen acted on the GnRH-primed pituitary to cause LH discharge, ovulation occurred, a corpus luteum was formed which regressed at the expected time, and the animals menstruated. The monthly cycles in these animals were thus determined by the ovaries, in the face of a constant, unvarying stimulus to the pituitary from the 'artificial hypothalamus'. However, it is important to realize that these experiments only demonstrate what *can* happen experimentally. It is manifestly not what *does* happen in life; hypothalamic pulsatility changes greatly during the cycle (see also Chapter 5).

When follicles begin to become scarce in the ovaries, as in women at the approach of the menopause, or in mules, which are born with a severely depleted stock of oocytes in their ovaries as a result of meiotic arrest in fetal life, then the cycles become irregular and abnormally long (see Fig. 6.3). This is presumably because the cycle is held up until one of the few remaining follicles in the ovary matures to the stage at which it becomes gonadotrophin sensitive.

Although all this evidence points strongly to the ovary as the 'clock' or 'zeitgeber' determining cycle length, there is equally compelling evidence to show that the brain can have important cyclical inputs. The frequency of pulsatile discharges of LH increases during the follicular phase to reach a maximum at the time of the pre-ovulatory surge, and declines during the luteal phase both in sheep and women. Laboratory rodents, and maybe even women, can only produce a pre-ovulatory LH surge in response to oestrogen at a set time of day. It is also very difficult to explain why the length of anovular menstrual cycles in primates soon after the menarche, when there is no corpus luteum present, or during early pregnancy when the corpus luteum is fully functional, is similar to the length of normal ovular cycles. And if we consider monoestrous species, like the dog or the roe deer (see Table 6.1), it is very difficult to imagine how the ovary could be the zeitgeber. Surely the brain must always have the last word when deciding on the timing of reproduction?

A true cycle has no beginning and no ending. All we can say is that the hypothalamus appears to be the spring that drives the reproductive clock (Chapter 1); much of the timekeeping of cycle length resides in the ovarian escapement. However, the clock will stop if not kept constantly wound up, and there are many circadian and circannual inputs to the hypothalamus that regulate the winding mechanism.

Puberty

The old explanation for the delay of cyclical ovarian activity until puberty was that immediately after birth, the hypothalamus and pituitary were thought to be extremely sensitive to the inhibitory effects of gonadal steroids; the ovary only had to produce a minute amount of oestradiol to inhibit gonadotrophin release. Puberty was viewed as a progressive *loss* of hypothalamic sensitivity to this feedback inhibition, thereby allowing increasing amounts of gonadotrophin to be secreted in the face of increas-

Fig. 6.16. Waves of follicular development followed by ovulation, formation and regression of the corpus luteum, and menstruation (M) in a 13-month-old pre-pubertal rhesus monkey pulsed once an hour with 6 μg GnRH for 110 days. (From E. Knobil. *Recent Progr. Hormone Res.* **36**, 53–88 (1980).)

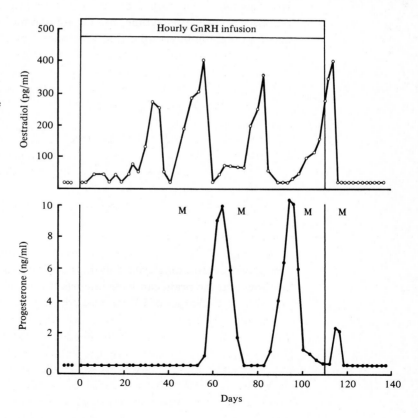

ing amounts of ovarian steroids. Thus everything could be explained by a resetting of the 'gonadostat' in the brain. We are now beginning to realize that this explanation is almost certainly incorrect. If ewes or female rhesus monkeys are castrated before puberty, or if girls with premature ovarian failure, as in Turner's syndrome, are followed through their pre-pubertal years, it can be shown that the immediate post-castration rise in gonado-trophins fails to occur. Thus it cannot simply have been gonadal steroid feedback that was holding gonadotrophin levels in check; there must be some steroid-independent inhibition of GnRH secretion in the pre-pubertal years. With the approach of puberty, there is an increasing frequency of pulsatile GnRH discharge from the hypothalamus (Chapter 1), so that the ovary or testis becomes progressively more and more stimulated. Ernst Knobil has shown that if pre-pubertal rhesus monkeys are given frequent pulsatile infusions of GnRH, ovulatory menstrual cycles can be induced, and the animals will revert back to a pre-pubertal state once the infusions are stopped (Fig. 6.16). Sequential blood samples taken from girls or boys going through puberty show that adult-type pulsatile LH secretion, and hence by inference hypothalamic GnRH secretion, begins during the

Fig. 6.17. Sequence of pubertal changes in girls and boys. (Adapted from W. A. Marshall. *J. Biosoc. Sci.* Suppl. **2**, 31–41 (1970).)

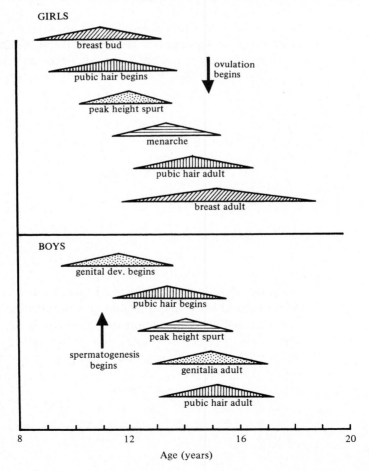

night-time, gradually extending to occupy the whole 24 hours – further evidence that circadian rhythms may be important in the regulation of human reproduction. However, the factors that ultimately determine why GnRH pulsatility should increase at puberty still remain unknown.

Puberty can be considered as either a state, or a stage of development. The Oxford Dictionary defines it as 'The state or condition of having become functionally capable of procreating offspring', and in animals we generally refer to puberty as that moment at which the female first comes into oestrus and ovulates, or the male first produces spermatozoa in his ejaculate. However in humans such criteria are almost impossible to determine; it requires frequent hormone assays to establish when a girl first ovulates, and it is not ethically possible to collect ejaculates from young boys. Human puberty is therefore best regarded as a sequence of developmental events occurring over several years, and Jim Tanner in London has done much to develop standard indices of pubertal development for girls and boys, based on changes in the breasts, pubic hair, testicular size, penis development, and growth in height and weight. It is interesting to compare the sequence of pubertal events in girls and boys (Fig. 6.17).

The first sign of impending puberty in a girl is breast enlargement, since the stromal tissue of the human breast is the most oestrogen-sensitive tissue in the body. This change is in response to the increasing waves of follicular development and oestrogen secretion from the ovaries (see Fig. 6.4), brought about by the nocturnal episodic discharges of LH, and a rising FSH secretion. The ovaries also begin to secrete androgens, and these, together with adrenal androgens, are responsible for the development of pubic and axillary hair. The next event is a sudden acceleration in the growth rate (Fig. 6.18), and this appears to be due to a stimulatory effect of small amounts of oestrogen on the rate of cell division in the epiphyseal cartilages of the long bones. However, this growth spurt is relatively short-lived, because as the oestrogen levels continue to rise they ultimately produce epiphyseal fusion and a cessation of long bone growth. At about the time that the growth rate is slowing down, the elevated oestrogen levels begin to stimulate the development of the uterine endometrium. This eventually becomes quite hypertrophied, so that when the oestrogen-secreting follicle becomes atretic, and the hormone levels fall, the endometrium is sloughed off and the first menstrual bleed, or menarche, occurs. As we have already seen, this may precede the first ovulation by a year or more, and the explanation usually given is that maturation of the positive feedback between oestrogen secretion and LH discharge is delayed (see Fig. 6.2). However, an even simpler and more plausible explanation is that the pituitary cannot discharge LH in response to oestrogen until it has been maximally primed by GnRH. At the time of menarche, this priming is still below the threshold necessary to elicit the discharge. Support for this view comes from Knobil's monkey experiments; when his pre-pubertal females were pulsed with optimal doses of GnRH, they all ovulated normally, and failed to show anovular cycles.

In boys, the sequence of pubertal changes is somewhat different from that seen in girls. The first noticeable change is an enlargement of the testes. This is because the diameter of the seminiferous tubules increases as spermatogenesis begins. Spermatogenesis is the result of a rising FSH secretion, which stimulates the Sertoli cells, and an increase in testosterone secretion from the Leydig cells in response to the nocturnal episodic discharges of LH. Only minute amounts of testosterone are produced initially, and although they may be inadequate to raise the concentration sufficiently in the systemic circulation to stimulate the development of the secondary sexual characteristics, there is enough to diffuse locally from the Leydig cells to the Sertoli cells and thereby stimulate meiosis in the spermatogonia embedded within them (see Chapter 4). Spermatozoa can be detected in the urine of young boys at the very beginning of pubertal development, even before the first pubic hair has appeared.

As testosterone levels continue to rise with increasing LH secretion, secondary sexual characteristics begin to develop. The penis grows, pubic and axillary hair appear, and testosterone stimulates cell division in the epiphyseal cartilages of the long bones, producing a growth spurt. The fact that this occurs so much later in boys than in girls (see Fig. 6.18) is probably due to the greater sensitivity of the epiphyses to oestrogen, both for stimulating growth and for bringing about epiphyseal fusion. One

Fig. 6.18. Height velocity curves in pubertal girls and boys. (From J. M. Tanner. *Fetus into Man*. Open Books; London (1978).)

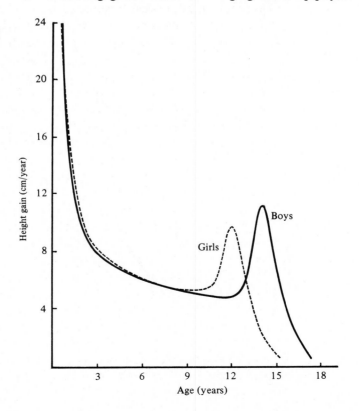

consequence of this is that parents often become worried when a younger sister begins to outgrow her older brother. But the fact that the boy starts his growth spurt later, and it is usually greater, means that boys almost invariably end up taller than their sisters. Androgens will also have other effects; they stimulate the development of striated muscle in boys, leading to greater muscular strength (Fig 6.19), and they produce changes in the shape of the larynx which result in breaking of the voice at around the age of 13.

Humans go through puberty later than any other mammal, and hence the sequence of pubertal changes is particularly long-drawn-out. In most animals, these changes are telescoped into a few months, weeks or days, so that it is difficult to dissect them apart. The pubertal growth spurt, for example, is too transitory to be recorded, except in other slow-developing species like elephants. Neoteny – the retention of infantile characteristics into adult life, and the postponement of puberty, may have been a particularly significant reproductive strategy adopted by man, since it prolongs the period of childhood dependency, thus affording more opportunities for the transmission of experience from one generation to the next. It is also interesting to see how the end result of these pubertal changes is that the woman becomes physically attractive to men some time before she becomes fertile, whereas the man becomes fertile long before he has acquired the physical attributes that will allow him to be successful in competition with older males. It is no accident that women in all societies usually marry men who are older than themselves.

Extraneous factors influencing cycles

Seasonal breeding

Many of the world's mammals live in the temperate regions of the globe, where seasonal changes in day-length result in large variations in food availability during the year: these become particularly pronounced at high

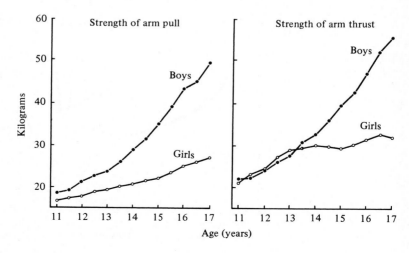

Fig. 6.19. Development of muscular strength in pubertal girls and boys. (From J. M. Tanner *Growth at Adolescence.* Blackwell; Oxford (1962).)

latitudes, and high altitudes. For the animals living in these areas, it therefore becomes critically important for the mother to give birth to her young in due season, which is usually the summer. Since the duration of gestation is relatively fixed (embryonic diapause being a notable exception; see Book 2, Chapter 2, Second Edition), this means that such species have had to regulate the timing of conception in order to control the season of birth. The small mammals, with short gestation lengths, can manage both to mate and to give birth during a summer, and so we refer to them as 'long-day breeders'; the golden hamster is a good example (see Table 6.1). However, for larger animals with longer gestation periods, it may become necessary to mate during the autumn or winter months of one year in order to ensure a summer birth in the next, and such animals are referred to as 'short-day breeders'; the sheep, goat and red deer are good examples. In species with gestation periods approaching a year, like the horse, or the roe deer, mating in the summer once more becomes the rule.

Species that have evolved in the tropics, where seasonal variations in food availability are much less pronounced, and less predictable, often lack any seasonality in the distribution of their births; chimpanzees and gorillas in Central Africa, and orang-utans in Borneo and Sumatra, are good examples of this. The fact that there is little or no seasonality in human births is perhaps a pointer to our own tropical origins. Where seasonality has been recorded in human populations, it often seems to be correlated with ambient temperature; it can just get too hot to copulate (see Fig. 6.20).

We can obtain some interesting information on the evolution of seasonal reproduction by studying the breeding seasons in closely related species distributed over a wide range of latitudes. The macaque monkeys are a good case in point; the bonnet macaque (*Macaca radiata*) of Southern India is apparently completely aseasonal, whereas the closely related rhesus monkey (*M. mulatta*) of Northern India and Southern China is a short-day breeder. The Japanese macaque (*M. fuscata*), which lives further from the equator than any other non-human primate, is again a short-day breeder, and it has the most restricted breeding season of all primates.

Fig. 6.20. Month of conception and ambient temperatures in Hong Kong. (From K. S. F. Chang *et al. Human Biol.* **35**, 366–76 (1963).)

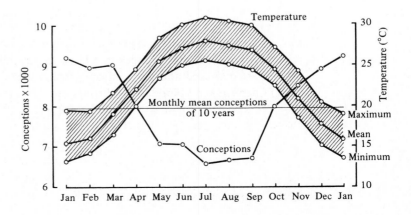

Merely bringing a tropical species to a temperate region of the world does not mean that it will instantaneously adopt a seasonal pattern of reproduction. The Barbary sheep of North West Africa (*Ammotragus lervia*), which breeds throughout the year in its natural habitat, has continued to remain aseasonal for many generations in the cold confines of Regent's Park Zoo in London; it pays a heavy price, since the lambs born in winter usually fail to survive. Similarly the Chital or Axis deer of India (*Axis axis*), perhaps the most beautiful of all the deer, remains aseasonal when transported to temperate regions of the world.

It would be a mistake to assume that all tropical species are automatically aseasonal. Some of the tropical bats, for example, have retained a distinct seasonality in the timing of birth, and this can be shown to be correlated with seasonal changes in rainfall, which in turn regulate the abundance of the insects on which the bats feed. Thus photoperiod, temperature and nutrition can all influence the seasonality of reproduction, but for temperate-zone species, photoperiod is by far the most important mechanism.

How is seasonal breeding controlled? Since female mammals have a far greater energy investment in reproduction than males – compare the energy demands of gestation plus lactation, with those of an ejaculation – it follows that the female is the limiting resource, and therefore she is the one who is principally concerned with regulating the matings. Although the male may also show seasonal changes in his reproductive activity, these are not as constrained as those of the female, so the male is potentially fertile for a longer period of the year.

Seasonal breeding has been likened to a recurrent annual puberty, and it is certainly true that the end result is the same; the seasonally quiescent animal, like a pre-pubertal one, has relatively inactive gonads as a result of a decreased frequency of GnRH secretion from the hypothalamus, and hence a decreased secretion of FSH and LH from the pituitary. As Gerald Lincoln has discussed in Chapter 3, the pineal gland plays a key role in regulating an animal's response to photoperiod, and hence the timing of its seasonal reproductive activity. However, the effects of the pineal on the timing of puberty are less apparent.

Fred Karsch and his colleagues have done much to unravel the precise endocrine mechanisms by which seasonal breeding is controlled in the ewe. In an elegant series of experiments they were able to show that there were clear-cut seasonal changes in LH concentrations in ewes that had been ovariectomized and given silastic implants of oestradiol-17β which maintained constant blood levels throughout the year; these changes were not seen in castrates without an oestradiol implant (see Fig. 6.21). They interpreted these results in terms of the 'gonadostat' model, and concluded that seasonal reproductive quiescence in the ewe is due to the fact that the changing photoperiod has in some way made the pulse generator in the hypothalamus extremely sensitive to the negative feedback effects of oestradiol during the long days of summer. As soon as a summer follicle starts to develop in the ovary and to secrete oestradiol, this feeds back onto the hypothalamus to inhibit the pulsatile secretion of GnRH, thereby suppressing LH secretion by the pituitary and inhibiting any further follicular development.

This neat explanation may not necessarily be the correct solution. Careful examination of the patterns of LH secretion in castrated ewes without any hormone replacement therapy shows that they still undergo some seasonal changes in the amplitude and frequency of their LH discharges, so photoperiod can exert a direct effect on the hypothalamic GnRH pulse generator in an 'open loop feedback' situation when there is no inhibitory steroid present. Nobody doubts that negative feedback is important in holding gonadotrophin levels in check; witness the spectacular rise in gonadotrophin levels following castration. But just as we have become suspicious of the gonadostat model to explain puberty, so must we be cautious if we are to resurrect it to explain seasonal breeding. Gerald Lincoln, who has studied the hormonal control of seasonal breeding in rams, is of the opinion that direct effects of photoperiod on the pulse generator are more important than indirect effects via changes in negative feedback sensitivity. The bicycle analogy may help to explain this conceptual difference in thinking between Fred Karsch and Gerald Lincoln. Consider the pedals of the bicycle as representing the hypothalamic pulse generator, and the brakes the inhibitory effect of gonadal steroids. The speed of the bicycle can be determined by the energy applied to the pedals, or the force applied to the brakes, or a combination of the two. Fred would argue that photoperiod operates predominantly by controlling the brakes, whereas Gerald would say that photoperiod primarily regulates the energy of

Fig. 6.21. Changes in LH concentrations throughout the year in ovariectomized ewes, and ovariectomized ewes implanted with oestradiol-17β in silastic so as to maintain a constant blood level of the hormone throughout the year. Note that the implanted animals have undetectable levels of LH during the period of summer anoestrus. No seasonal changes are apparent in the unimplanted ovariectomized controls; the small changes that do occur are masked by plotting the data on a logarithmic scale, and by infrequent blood sampling that is unable to detect alterations in pulsatility. (From F. J. Karsch. *Physiologist*, **23**, 29–38 (1980).)

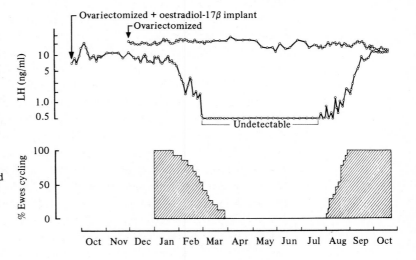

pedalling. Thus his interpretation of Fred's experiment in Fig. 6.21 is that it merely reveals the photoperiodically induced changes in the activity of the hypothalamic pulse generator in the presence of a constantly applied steroidal inhibition, and does not necessarily tell us anything about negative feedback sensitivity. The truth probably lies somewhere between these two extremes, but it is going to be extremely difficult to design experiments that tease apart hypothalamic drive and negative feedback.

Changes in photoperiod control not only hypothalamic GnRH secretion, but also the ability of the animal to respond behaviourally to a given sex hormone. Progesterone-primed, ovariectomized ewes will show oestrus much more readily in response to administered oestrogen during the normal mating season than during summer anoestrus, and castrated red deer stags will show rutting behaviour in response to testosterone only during the normal rutting season in the autumn. Even ovariectomized canaries will build nests in response to oestrogen treatment only in the spring. The effects of photoperiod are thus far-reaching, and we have much to discover. It will be fascinating to see how many of the photoperiodic controls of reproductive events are pineally mediated, and precisely how melatonin acts to regulate hypothalamic activity.

Lactation

Lactation shares with seasonal breeding the distinction of being one of nature's principal contraceptives. It has relatively little impact on fertility in small mammals, which are geared to high rates of reproduction, although it can delay the next birth by a few weeks even in rats and mice by inducing a transitory state of embryonic diapause. In the larger seasonally breeding mammals, the inhibitory effects of lactation often merge into and complement seasonal reproductive quiescence, to bring about an annual spacing of births; sheep and Tammar wallabies provide two rather different examples of how this duality of inhibitory effects operates.

Ewes first come into oestrus in the autumn (August to November, depending on the breed), and if they are not mated, they will continue to have regular 16-day oestrous cycles until February or March of the following year. Thus a ewe that was mated in August would lamb in January the following year. If there was no lactational inhibition of oestrus, such an animal might be able to conceive again in February or March, to give birth to a second set of lambs in July or August. However, the suckling of the first set of lambs provides contraceptive cover for the remaining months of the ewe's mating season, so that a second pregnancy is not normally possible during that time. By the time that the lambs are weaned in the early summer, lactational anoestrus has given way to seasonal anoestrus once more.

The Tammar wallaby shows another variation on this theme (see also Book 2, Chapter 2, Second Edition). Birth of the immature joey occurs about a month after the summer solstice (longest day), and within a few hours of birth the mother has returned to oestrus, ovulated, and conceived again. But the suckling activities of the joey, by now permanently attached to one of the four teats in the pouch, causes an increased secretion of prolactin from the pituitary; this then inhibits progesterone production by the newly formed corpus luteum, suppresses endometrial development, and holds the new blastocyst in a state of embryonic diapause. This lactational inhibition persists for about six months, and if the pouch young is removed during this time, the corpus luteum will be reactivated and the arrested blastocyst will resume its development. After six months, lactational diapause gives way to seasonal diapause, when removal of the pouch young no longer has any effect on the blastocyst. Seasonal diapause is terminated at the time of the summer solstice, when the corpus luteum reactivates, the blastocyst resumes development, and birth occurs about four weeks later, shortly followed by a new ovulation and conception. Thus the Tammar wallaby only spends a few hours a year when it is not pregnant, but since 11 months are spent in embryonic diapause, it is a painless compromise.

The most dramatic inhibitory effects of lactation on ovulation are seen amongst the higher primates, including humans. As we have already mentioned, in these long-lived species with slow rates of postnatal development, low fertility has been at a premium. Photoperiod cannot be used to inhibit reproduction for longer than a year at a time, and so this is where lactational inhibition comes into its own. The small primates, like the marmoset monkey, are still geared to high rates of reproduction; they usually give birth to twins, or even triplets, and there is no evidence that lactation has any inhibitory effect on ovulation whatsoever (see Table 6.2). The baboons have evolved a moderate degree of lactational inhibition, but it is man (exemplified by the !Kung hunter–gatherers) and the great apes that have developed the mechanism to its fullest extent, so that birth intervals of four or five years are the norm. One great advantage of lactational inhibition over seasonal inhibition is that if the infant dies,

Table 6.2. *Birth intervals in primates living in the wild in relation to gestation length*

	Common marmoset	Yellow baboon	Olive baboon	Gorilla	Chimpanzee	!Kung hunter–gatherer
Gestation length (months)	5	6	6	8.5	8	9
Mean birth interval (months)	6.5	23	26	46	68	49

lactation will immediately cease, ovulation will recommence, and the mother will soon become pregnant again.

The precise mechanisms by which lactation can inhibit ovulation are still poorly understood. Afferent neural impulses from the teat are essential, since denervation of the teat abolishes all the inhibitory effects of suckling. The temporal distribution of the suckling bouts themselves is also of critical importance; it has recently been shown in beef cattle that if calves are only allowed to suckle their mothers for 30 minutes a day, starting on the 30th day of life, their mothers return to oestrus about 100 days sooner than if the calves had been allowed to run with their mothers continuously and suckle at will (Fig. 6.22). Along the same lines, Peter Howie and Alan McNeilly in Edinburgh have shown in women that the early abandonment of night-time feeding, coupled with a much reduced suckling frequency during the day-time, has been a significant factor in reducing the contraceptive effect of breast-feeding.

When afferent neural stimuli from the teat reach the hypothalamus, they inhibit the release of dopamine into the pituitary portal circulation, thereby giving rise to an increased secretion of prolactin from the anterior pituitary. As we have seen in marsupials such as the Tammar wallaby, this hyperprolactinaemia results in reduced activity of the corpus luteum, and the arrest of embryonic development. In eutherian mammals, it is increasingly doubtful whether hyperprolactinaemia is responsible for the inhibition of ovulation; instead, there is growing evidence to suggest that the afferent neural inputs may increase hypothalamic production of β-endorphin, one of the opioid peptides produced by the brain, and that this in turn may be responsible for depressing GnRH secretion, and so inhibiting ovulation. Certainly, women who are hyperprolactinaemic and amenorrhoeic as a result of prolactin-secreting pituitary tumours can have

Fig. 6.22. The effect of reduced suckling time on the duration of lactational anoestrus in beef cows. Starting on the 30th day post-partum, one group of calves were only allowed to suckle their dams for 30 minutes a day, whereas the control group were run with their dams continuously, and allowed to suckle at will. (From R. D. Randel. *J. Anim. Sci.* **53**, 755–7 (1981).)

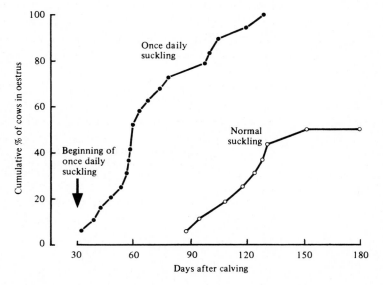

their pattern of LH secretion returned to normal by administration of the opiate antagonist naloxone, which has no effect on the hyperprolactinaemia.

Nutrition

There is abundant evidence in a variety of laboratory, domestic and wild animals to show that an extremely low plane of nutrition, bordering on starvation, will arrest all reproductive activity. A low plane of nutrition after birth will also delay the onset of puberty, and it is generally agreed that the decline in the age of menarche experienced in many Western countries in the last century was due to improvements in nutritional status during infancy and childhood. In the Bundi tribe, ekeing out a meagre existence in the highlands of New Guinea, the average age at menarche today is 18, and there is little doubt that this would soon approach the Western norm of 12–13 if the diet could be improved. This much is fact, but there has been a great debate as to the precise mechanisms by which nutrition exerts its effect on reproduction. Rose Frisch in Harvard has championed the view of a 'critical body weight' for human menarche of 47 kg, and she believes that this reflects a critical percentage of body fat. Although the idea is appealing, particularly since we know that body fat is an important site for the aromatization to oestrogens of circulating androgens of adrenal or ovarian origin, it has been severely criticized. There is bound to be some relationship between accelerated growth in childhood and earlier sexual maturity, but to say that a critical weight, or a critical mass of fat, is the trigger for the onset of puberty is to invoke causality where none may exist. The debate becomes even more heated when we consider the effects of nutrition on the duration of lactational amenorrhoea. Rose Frisch claims that nutrition may be the principal factor determining the long periods of amenorrhoea seen in some human communities. However, John Bongaarts at the Population Council in New York has convincing evidence to show that plane of nutrition is responsible for only about 10 per cent of the variability in length of post-partum amenorrhoea seen in different communities, and that the major factor is undoubtedly the suckling stimulus itself.

The subtle nature of the interactions between nutrition and lactation has been beautifully illustrated by Andrew Loudon and his colleagues, working on wild red deer in Scotland. Hinds on poor hill grazings produced much less milk than those on better pastures; as a result, their calves were hungrier, suckled much more frequently, and grew more slowly, whilst their mothers lost condition. Since suckling frequency is one of the keys to the contraceptive effect of lactation, it is easy to see how the effects of nutrition and lactation could be compounded.

Olfaction and social factors

We have already referred to the importance of pheromones for providing evidence of oestrus in elephants, red deer, cows, pigs, and dogs, but these olfactory hormones are assuming a wider significance the more we study them. Farmers have long since known that the introduction of a ram into a flock of ewes just before the mating season will hasten its onset, and provide some degree of oestrous synchrony. As Fred Karsch points out in the opening chapter, we now know the reason for this: the presence or even just the smell, of the ram produces an increased frequency of pulsatile LH discharge in the ewes, thereby accelerating follicular development and ovulation. A somewhat similar phenomenon is seen in pigs, where early introduction of a boar into a group of sows will significantly hasten the onset of puberty (see Fig. 6.23). It is interesting to note that if the olfactory bulb of the brain is surgically removed from sows, they lapse into a state of anoestrus. Male pheromones, or rather the lack of them, also seem to be responsible for the well-known effects of different social conditions on the oestrous cycles of mice and rats. If female mice are caged individually in the absence of males, they show a high incidence of extended dioestrous cycles of 7 days or more; if they are caged in groups, the cycles become even longer due to the spontaneous occurrence of pseudopregnancies. The introduction of a male mouse into a group of females will bring many of the females into oestrus 3 days later, and this synchrony can even be produced just by adding male urine. Male urine will also hasten the onset of oestrus in a group of female rats or guinea pigs. The pheromone in mouse urine, whatever it may be, is certainly androgen-dependent, since the urine of castrated males is ineffective,

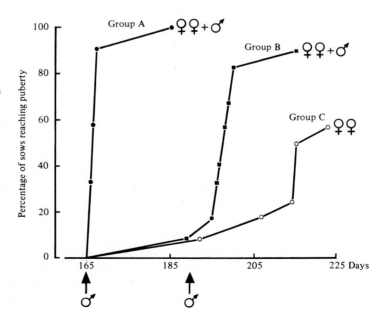

Fig. 6.23. The effect of an adult boar on the time of onset of first oestrus (puberty) in a group of sows. In Group A, the sows were exposed to several different boars from day 165 onwards. In group B, a pair of boars was introduced on day 190. Group C was kept in isolation from boars throughout the experiment. (From P. H. Brooks and D. J. A. Cole, *J. Reprod. Fert.* **23**, 435–40 (1970).)

whereas urine from androgenized females is indistinguishable from that of normal males. The human nose is readily able to detect differences in the smell of male mouse urine from that of intact or castrated females or castrated males. And anybody who has kept a tomcat will have noticed how castration improves the smell of the entire neighbourhood.

The time of onset of puberty can also be hastened in mice and rats by the presence of a male, or of male odours. But perhaps one of the most unexpected effects of pheromones was the discovery by the late Hilda Bruce that the smell of a strange male mouse could terminate pregnancy in 70–80 per cent of newly mated females; however, by the 6th day of pregnancy, the females had become immune to this abortifacient effect.

There can be no doubt that pheromones exert an important 'fine-tuning' control on reproductive rhythms in a wide variety of species, and much work remains to be done to define the chemical nature of the substances concerned, their mode of detection – whether by the olfactory epithelium, or the vomeronasal organ, or both – and their central mechanism of action within the brain itself.

Some of the most surprising social effects on cycle lengths are seen in our own species, although whether these have any pheromonal basis seems doubtful. Psychogenic amenorrhoea came into prominence during the second world war, when wives were often separated from their husbands for long periods of time under distressing circumstances. One study showed that 14.8 per cent of British and American women in a Japanese internment camp in the Phillipines became amenorrhoeic, often for a year or more, even though they tended to gain weight during their captivity. In another study, it was shown that a severe emotional disturbance, like the break-up of a romance, or the death of a close relative, could postpone ovulation and menstruation if the precipitating event occurred in the follicular phase of the cycle, but the effects were much less pronounced if the event occurred in the luteal phase; this means the life-span of the corpus luteum was evidently unaffected, and the individual was given time to adjust to the circumstances before the onset of the next follicular phase. We have already referred to the fact that 'jet lag' in the follicular phase likewise seems to be particularly disturbing to cycle length, presumably by inhibiting ovulation. Rose Frisch and her colleagues have also shown that college athletes in training, be they runners or swimmers, may have their menarche postponed by two or three years through some unknown mechanism.

One of the most intriguing studies of a social effect on cyclical activity was Martha McClintock's report that students living in a residential women's college in the USA appeared to synchronize their menstrual cycles with those of their room-mates; there was also a suggestion that girls who frequently 'dated' boys had shorter menstrual cycles. This menstrual synchrony effect has recently been confirmed in a study of 79 students at the University of Stirling in Scotland, although the investigators could find no effect of the amount or nature of social interactions with males on cycle lengths. As in the McClintock study, it appeared to be the amount of time that two individuals spent together that determined whether their cycles became synchronous, even though in most instances the girls were unaware of the timing of their close friend's cycles.

Female mammals have used every stratagem to ensure that mating takes place at the time of ovulation. During oestrus, the female becomes attractive to the male in terms of her appearance, her smell, her behaviour, and even her vocalizations; she will exhibit proceptive behaviour, often wandering over a considerable distance to seek out the male, and once the two have met she will become receptive to his advances, and will facilitate intromission by positioning herself accordingly. So far as we know, ours is the only species in which oestrous cycles do not occur, thereby providing fertile ground for much sociobiological speculation. Endlessly repeated cyclical activity is abnormal, and may be too much of a good thing; there are well-recognized hazards of nulliparity in women as far as the ovaries, the uterus and the breasts are concerned. These adverse effects could probably be overcome by the judicious design of new contraceptives.

Puberty is the great awakening, when the whole reproductive tract comes to life. Ideas are changing as to how this is brought about; we can document the details, but the precipitating event continues to elude us. Once reproductive activity has become established, nature has designed a whole array of checks and balances to regulate fertility, and these operate principally in the female who has the greater energy investment in reproduction and is therefore the limiting resource. Nature's two principal contraceptives are seasonal breeding – designed to ensure that the young are born at a time of year that maximizes their chance of survival – and the lactational inhibition of ovulation – designed to ensure that a new birth does not follow hard on the heels of the preceding one. They involve very different inputs, namely daylight length and suckling frequency, but the end result is the same: the secretion of GnRH is suppressed and ovulation fails to occur. Olfactory cues are frequently used for the fine-tuning of reproductive events, and social factors may also be important in certain circumstances, although the mechanisms by which they act are quite unknown.

The vast array of different reproductive strategies displayed by female mammals is testimony to the variety of ecological niches that mammals have successfully colonized. In order to make sense out of this bewildering mass of information, we must go back to the wild, and begin to understand the circumstances under which the reproductive mechanisms first evolved.

Suggested further reading

Mechanisms of control of the reproductive function by olfactory stimuli in female mammals. C. Y. Aron. *Physiological Reviews*, **59**, 229–84 (1979).
Does malnutrition affect fecundity? A summary of evidence. J. Bongaarts. *Science*, **208**, 564–9 (1980).

Physiological problems of seasonal breeding in Eutherian mammals.
  J. R. Clarke. In *Oxford Reviews of Reproductive Biology*, pp. 244–312. Ed.
  C. A. Finn. Clarendon Press; Oxford (1981).
The role of the vomeronasal organ in mammalian reproduction. R. D. Estes.
  Mammalia, **36**, 315–41 (1972).
Population, food intake and fertility. R. E. Frisch. *Science*, **199**, 22–30 (1978).
Malnutrition and fertility. R. E. Frisch. *Science*, **215**, 1272–3 (1982).
Menstrual synchrony in female undergraduates living on a coeducational
  campus. C. A. Graham and W. C. McGrew. *Psychoneuroendocrinology*, **5**,
  245–52 (1980).
Seasonal reproduction: a saga of reversible fertility. F. J. Karsch. *The
  Physiologist*, **23**, 29–38.
The neuroendocrine control of the menstrual cycle. E. Knobil. *Recent Progress
  in Hormone Research*, **36**, 53–88 (1980).
Induced ovulation in mammals. S. R. Milligan. *Oxford Reviews of Reproductive
  Biology*, pp. 1–46. Ed. C. A. Finn. Clarendon Press; Oxford (1982).
Asian bull elephants: flehmen-like response to extractable components in female
  elephant estrous urine. L. E. Rasmussen, M. J. Schmidt, R. Henneous,
  D. Groves and G. D. Daves. *Science*, **217**, 159–62 (1982).
The evolution of human reproduction. R. V. Short. *Proceedings of the Royal
  Society of London, Series B*, **195**, 3–24 (1976).
The discovery of the ovaries. R. V. Short. In *The Ovary, volume I. General
  Aspects*. Ed. Lord Zuckerman and Barbara J. Weir. Academic Press; London
  (1977).
Menstruation. S. T. Shaw and P. .C. Roche. *Oxford Reviews of Reproductive
  Biology*, vol. 2, pp. 41–96. Ed. C. A. Finn. Clarendon Press; Oxford (1980).
Photoperiodism and Reproduction in Vertebrates. Ed. R. Ortavant, J. Pelletier
  and J.-P. Ravault. INRA Colloquium **6**. INRA; Paris (1981).
Pre-menstrual tension. E. M. Symonds. *Oxford Reviews of Reproductive Biology*,
  vol. 3, pp. 156–81. Ed. C. A. Finn. Clarendon Press; Oxford (1981).
Biological Basis for the Contraceptive Effects of Breast Feeding. R. V. Short. In
  Advances in International Maternal and Child Health, vol. 3, ed. D. B. Jelliffe
  & E. P. Jelliffe, pp. 29–39.
Patterns of Mammalian Reproduction. S. A. Asdell. Cornell University Press;
  New York (1964).
Reproductive Biology of the Great Apes. Ed. C. E. Graham. Academic Press;
  New York (1981).
Neuroendocrine Aspects of Reproduction. Ed. R. L. Norman. Academic Press;
  New York (1983).

7

Pregnancy

R. B. HEAP and A. P. F. FLINT

Hormones have important regulatory roles in pregnancy from the moment of ovulation and fertilization to the delivery of young; in this chapter we shall be concerned with how hormones regulate the maternal adjustments to pregnancy, and how their secretion is controlled to meet the needs of uterine gestation. When viviparity emerged as the preferred mode of reproduction in eutherian and marsupial mammals, hormones were exploited to control the physiological adjustments of the mother. The success of viviparity in eutherians and marsupials was due to several important features: reduction in the yolk content of the eggs, development of the placenta, retention of the young within the female genital tract, and parental care of the offspring, especially including their nutrition. Such functions presented problems for maternal homeostasis, and one of the ways by which these were solved involved the exploitation of pre-existing molecules (steroids and polypeptides) as regulators of organ function. The sex hormones and gonadotrophins of mammals are widely distributed in Nature and are also involved in reproduction in many non-mammalian species. A study of the role of hormones in pregnancy therefore provides an exciting opportunity to see the uses to which these conserved molecules have been put in the evolution of viviparity.

Endocrine regulation of pregnancy primarily involves hormones of the pituitary, the ovary, and the placenta, and a distinction can be drawn between these organs and those, such as thyroid, adrenal and parathyroid, whose role is more of a supportive or permissive one. Most hormones are released from the gland into the circulation and act at a distance from it, but some are produced within the organs where they act, and this group of local hormones has been closely studied in recent years. Certain hormones, such as human chorionic gonadotrophin and progesterone, have been implicated in the immunoregulation of pregnancy, enabling the mother to accommodate the fetal allograft; this topic is dealt with in Book 4, Chapter 6, First Edition.

Maternal recognition of pregnancy

How an animal knows that it is pregnant, stops having oestrous cycles and delays ovulation, and why it does not reject an immunologically foreign fetal 'allograft', are central problems arising from the adoption of viviparity. Recently, Hallie Morton and her colleagues in Brisbane have suggested that within a few hours after mating the embryo produces an immunosuppressive factor which may protect it initially from the maternal immune system. Early Pregnancy Factor, as this new agent has been called, has been detected in mice within 6 hours after mating by an *in vitro* reaction, the rosette-inhibition test. This *in vitro* test depends on the ability of T-lymphocytes to form rosettes with heterologous red blood cells. Rosette formation is inhibited by anti-lymphocyte serum, but, if the lymphocytes have been exposed previously to an immunosuppressive agent, much less antilymphocyte serum is required to inhibit rosette formation. Mouse early pregnancy factor can be heat-inactivated at 72 °C but not at 56 °C, is not dialysable, and has immunosuppressive properties as shown by the inhibition of delayed-type hypersensitivity. The factor is not limited to one species since it has been found in women, sheep and cows, and experiments in sheep show that its occurrence is not due to ovulation or to spermatozoa in the uterus, but to the presence of a live embryo. These findings have excited interest in several laboratories but they have been difficult to corroborate. If correct, the presence of an immunosuppressive factor at such an early stage must be the earliest indication of fertilization so far detected in the mother.

Maternal recognition of pregnancy is a phrase first coined by Roger Short in 1969; it has different meanings for different disciplines. For the immunologist it means maternal adjustments that allow the retention of a resident allogeneic embryo rather than its rejection as a foreign tissue; for the cell biologist, it refers to the morphological and physico-chemical interactions that occur at the surface of the early embryo and uterine epithelium. Prior to implantation, changes that occur in the glycocalyx covering the epithelium of the endometrium, or changes in the DNA- and protein-synthetic activity of presumptive sites of attachment, also show that the blastocyst has localized effects. For the endocrinologist, maternal recognition of pregnancy is reflected in the way that the regression of the corpus luteum is first prevented or arrested by the presence of a conceptus. The corpus luteum is a characteristic, though not a unique, feature of mammalian pregnancy which plays a specific role in the endocrinology of gestation through the synthesis and secretion of progesterone, the hormone of pregnancy.

In mammals the life of the corpus luteum, which is normally a transient organ, can be extended by one of several mechanisms. Of these, the best understood is the luteotrophic stimulus, in which pituitary luteotrophins are secreted at the time of mating (as in the rat). Another common mechanism is the antiluteolytic stimulus, by which the conceptus inhibits the normal luteolytic action of the uterus mediated by prostaglandin $F_{2\alpha}$ (as in the sheep). A third type of mechanism may involve a placental luteotrophin which also has an antiluteolytic action (as in man). The general idea that embryos produce some message after they have entered the uterus, but before they have implanted, thereby extending the life of the corpus luteum, has become well established. In women and rhesus monkeys, pregnancy is probably recognized through a luteotrophin secreted by the conceptus, which prolongs the life of the corpus luteum and delays menstruation (Fig. 7.1). Chorionic gonadotrophin is thought to be the placental luteotrophin in question; in primates it is produced from about the time of implantation, so that active or passive immunization of women or monkeys against this hormone (hCG or its β-subunit) will prevent the establishment of pregnancy and allow normal cycles to continue. The timing of the production of chorionic gonadotrophin is critical, since in rhesus monkeys the secretory activity of the corpus luteum

Fig. 7.1. How hormones produced in early pregnancy may prolong the life of the corpus luteum (CL) in women and rhesus monkeys. In the menstrual cycle the life of the corpus luteum is regulated locally by ovarian hormones such as oestradiol and prostaglandins. In pregnancy the embryo in the uterus produces a luteotrophin (chorionic gonadotrophin) which overrides the local luteolytic influence of the ovarian hormones, and prolongs the life of the corpus luteum.

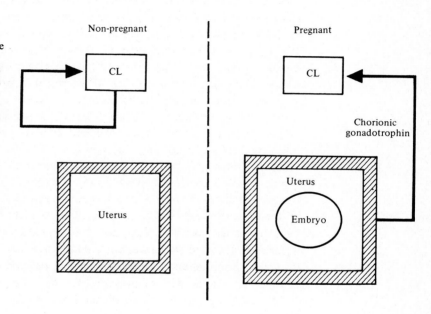

Fig. 7.2. The time when the corpus luteum is rescued from regression in women. The points denote the mean concentrations ($\pm$SD) of progesterone and hCG in the plasma of ten women who conceived spontaneously. The day when hCG was first detectable is designated day 1; this occurred 7 to 10 days after ovulation as inferred from the LH peak (E. A. Lenton, personal communication).

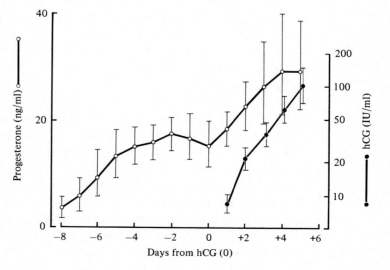

begins to decline but is then 'rescued' by the luteotrophic signal, as Ernst Knobil describes it, at almost the last possible moment. Direct evidence for a very similar form of corpus luteum rescue in women has been described recently by Elizabeth Lenton in Sheffield (Fig. 7.2). In sheep, and other domestic ruminants, the message is transmitted two or three days before the animal would normally return to oestrus, but in these species the conceptus prevents secretion of the uterine luteolysin, prostaglandin $F_{2\alpha}$ ($PGF_{2\alpha}$) (Fig. 7.3).

An exciting new finding which may help to explain how $PGF_{2\alpha}$ causes regression of the corpus luteum is the discovery that luteal cells secrete oxytocin. Claire Wathes and Ray Swann at Bristol have found that the sheep's corpus luteum contains almost as much oxytocin as does the neurohypophysis (the traditional source – see Chapter 2), and in our laboratory at Babraham we have found that when the corpora lutea normally regress at the end of the oestrous cycle, episodes of release of $PGF_{2\alpha}$ from the uterus are accompanied by surges of secretion of oxytocin from the corpus luteum. Moreover, the release of oxytocin by the sheep's ovary is greatly increased when a luteolytic prostaglandin analogue is administered (Fig. 7.4). Since isolated luteal cells in culture respond to oxytocin with reduced steroid secretion, luteal oxytocin may mediate or even augment the luteolytic effect of $PGF_{2\alpha}$. The discovery of oxytocin production by the corpus luteum, and the earlier finding that the ovary contains and responds to gonadotrophin releasing hormone (GnRH) (see Chapter 5), underlines the wide distribution of these biologically active small peptides and the variety of uses to which they have been put.

The presence of an agent in the sheep conceptus that neutralizes the luteolytic properties of the uterus (Fig. 7.3) was first recognized during the

Fig. 7.3. The conceptus prolongs the life of the corpus luteum in sheep by producing a protein, trophoblastin (TB), which reduces the release of a luteolysin, probably prostaglandin $F_{2\alpha}$ ($PGF_{2\alpha}$), by the uterus.

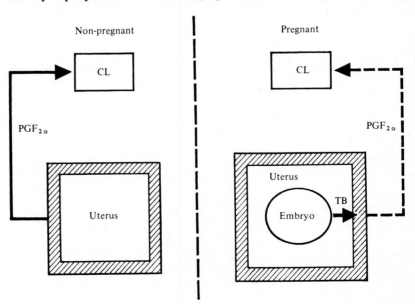

1960s by Bob Moor and Tim Rowson in Cambridge. Jacques Martal's group in France have now identified a heat-sensitive proteinaceous compound termed trophoblastin, which is produced for a few days before implantation and extends the life of the corpus luteum for more than one month.

Experiments by our group at Babraham and in Fuller Bazer's laboratory in Florida have shown another way by which the conceptus can produce a signal that neutralizes the luteolytic factor from the uterus, and prolong the life of the corpus luteum. In pigs, blastocysts undergo a dramatic elongation 11 to 12 days after fertilization or about one week before definitive attachment. Simultaneously, they begin to produce oestrogens which act locally to cause a redirection of the secretion of $PGF_{2\alpha}$ by the uterus, so that a much reduced quantity of this potent luteolytic agent is released into the uterine vein. The oestrogens themselves are enzymatically modified in the uterine wall (Fig. 7.5), but in circulation they may also act as luteotrophins (Fig. 7.6).

However, the life of the corpus luteum is not always prolonged. In marsupials, such as the Tammar wallaby and red kangaroo, the life-span of the corpus luteum in the pregnant and non-pregnant animal is similar, and the gestation period is about the same length as the oestrous cycle. There are some species of eutherian mammals like the ferret and dog that

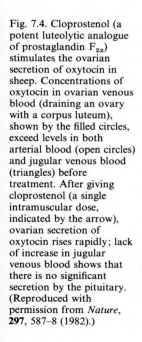

Fig. 7.4. Cloprostenol (a potent luteolytic analogue of prostaglandin $F_{2\alpha}$) stimulates the ovarian secretion of oxytocin in sheep. Concentrations of oxytocin in ovarian venous blood (draining an ovary with a corpus luteum), shown by the filled circles, exceed levels in both arterial blood (open circles) and jugular venous blood (triangles) before treatment. After giving cloprostenol (a single intramuscular dose, indicated by the arrow), ovarian secretion of oxytocin rises rapidly; lack of increase in jugular venous blood shows that there is no significant secretion by the pituitary. (Reproduced with permission from *Nature*, **297**, 587–8 (1982).)

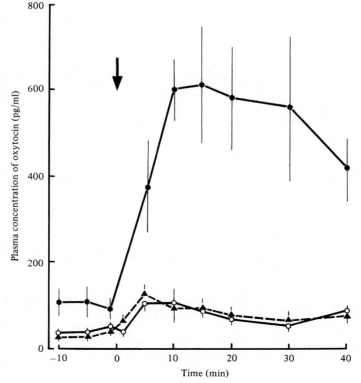

emulate the marsupials in having a luteal phase of similar length to the gestation period, so that 'pregnancy recognition' may not even be necessary. There are others in which the presence of a conceptus does not affect the function of the corpus luteum until much later in pregnancy; in the rabbit this occurs about one week after implantation.

Some species show a phenomenon known as delayed implantation (Book 2, Chapter 2, Second Edition), where the period of pre-implantation may exceed that of post-implantation. The delay may be caused by lactation, as in the rat, mouse, wallaby and kangaroo (so-called facultative delay), or by environmental changes, as in the badger or roe deer (obligatory delay). The delay found in lactating rats is related to the number of young suckled and does not usually occur if the litter size is less than five; implantation can be induced by a single injection of a minute amount of oestrogen. The delay found in the lactating tammar wallaby

Fig. 7.5. Production of oestrogens by the pre-implantation conceptus in pigs. Oestrone and oestradiol-17β are synthesized in the trophectoderm from maternal precursors and are then conjugated in the uterus to produce oestrone sulphate.

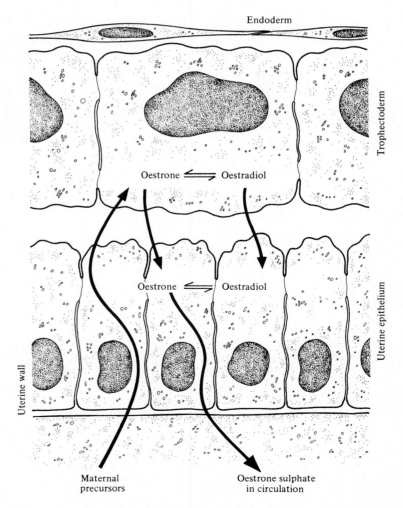

can be terminated by injections of the dopaminergic drug bromocriptine which allows reactivation of the corpus luteum, resumption of embryonic growth, and hence implantation. In the badger and other mustelids the termination of implantation is also associated with increased activity of the corpus luteum, but in the roe deer there is an increased secretion of oestrogens, probably derived from the conceptus.

The phenomenon of delayed implantation poses the question: when does the mother first respond to the presence of an embryo in the uterus? Is it when embryonic growth is reactivated after diapause, or does maternal recognition occur much earlier when periods of oestrous activity cease? A remarkable example of a precocious form of maternal recognition is that found in the horse, and was first described independently by Dr C. van Niekerk in South Africa and Professor Bielanski in Poland. They found that only *fertilized* horse eggs reach the uterus, the unfertilized eggs remaining trapped at the isthmus of the Fallopian tube where they often undergo parthenogenetic cleavage before slowly degenerating over the next few months (Fig. 7.7). In the event that another egg is ovulated at a later oestrus and subsequently fertilized, the developing morula can bypass its degenerating predecessors and enter the uterus. Whether this form of maternal recognition is achieved by differences in the surface properties of the two types of eggs remains a mystery, but this ability to differentiate between eggs precedes the time when the life of the corpus luteum is prolonged by the presence of an embryo by at least two weeks.

Faced with such a remarkable array of species differences, we are forced to conclude that there is no single mechanism for the recognition of pregnancy that is common to all species.

Fig. 7.6. The conceptus prolongs the life of the corpus luteum in pigs by producing oestrogen (OE) which acts as a luteotrophic hormone and also greatly reduces the release of uterine $PGF_{2\alpha}$ into the maternal circulation.

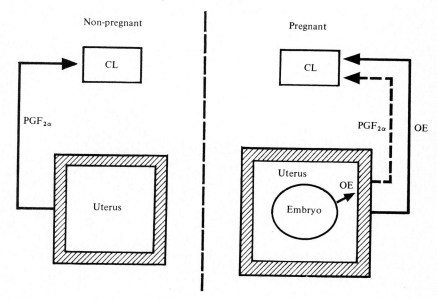

Hormone production by the ovary

We have already alluded to the fact that prolongation of the life of the corpus luteum is essential for the establishment of pregnancy in many mammals. The first indication of the important role of the corpora lutea in pregnancy maintenance came at the beginning of this century when Ludwig Fränkel demonstrated that removal of the corpora lutea from a pregnant rabbit terminated gestation. He was fortunate to choose the rabbit, for in this species ovariectomy always causes abortion; in others,

Fig. 7.7. Normal horse eggs before and after fertilization, and degenerating tubal eggs. (From C. H. van Niekerk and W. H. Gerneke. *Onderstepoort J. Vet. Res.* **33**, 195 (1966).)

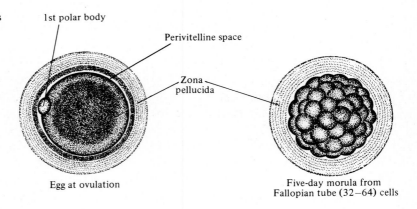

1st polar body

Perivitelline space

Zona pellucida

Egg at ovulation

Five-day morula from Fallopian tube (32–64) cells

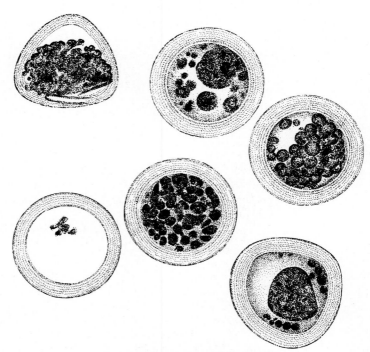

Degenerating ova from one Fallopian tube, ranging in age from 24 h –7½ months

Table 7.1. *Removal of the ovaries or pituitary in pregnancy and its effect on the maintenance of gestation in various species*

Animal and length of gestation (days)	Approximate stage of pregnancy when operation performed			
	Ovariectomy		Hypophysectomy	
	First half	Second half	First half	Second half
Woman (267)	+	+	+	+
Rhesus monkey (165)	+	+	+	+
Tammar wallaby (29)*	+	+	+	+
Quokka (27)*	+	+		
Guinea pig (68)	±	+	+	+
Sheep (148)	−	+	−	+
Horse (350)	−	+		
Brush-tailed possum (17)	−	+		
Rat (22)	−	±	−	+
Cat (63)	−	±		±
Ferret (42)	−	±	−	±
Cow (282)	−	±		
Dog (61)	−		−	±
Rabbit (28)	−	−	−	−
Mouse (19)	−	−	±	+
Hamster (16)	−	−	−	+
Goat (150)	−	−	−	−
Pig (113)	−	−	−	−
Virginia opossum (13)	−	−		
Armadillo (150)	Implantation − may occur			

+, Fetuses survive; ±, some fetuses survive; −, fetuses aborted or resorbed.
* From ovulation to birth, excluding the period of diapause.

Table 7.2. *Progesterone monoclonal antibody blocks pregnancy in mice*

Time of injection after mating	Group	No. pregnant/ no. treated	Total no. of implantations
32 h	Antibody	0/13	0
	Control	13/14	109
109–130 h	Antibody	2/11	20
	Control	7/12	75

Animals were injected with antibody before (at 32 h after mating) or during the time of implantation (at 109–130 h); autopsy between days 10 and 14 of pregnancy. (From L. J. Wright, A. Feinstein, R. B. Heap, J. C. Saunders, R. C. Bennett and M.-Y. Wang, *Nature*, **295**, 415–17 (1982).)

Fig. 7.8.
Electronmicrograph of a
luteal cell during early
pregnancy in the guinea
pig. There is an abundance
of smooth endoplasmic
reticulum typical of a
steroid-secreting cell, many
mitochondria, and
occasional lipid droplets
(ld) and lysosomes (ly).
× 14,800. Inset, the Golgi
complex. × 32,500.

such as the guinea pig, sheep or horse, pregnancy may continue if the operation is performed after a certain time in gestation (Table 7.1). So far as we know, there is no species in which pregnancy can be maintained in the total absence of progesterone. The important role of progesterone in early gestation can be shown by passive immunization against it. In mice, a monoclonal antibody raised against progesterone has been found to block implantation (Table 7.2). After immunization, circulating progesterone levels remain high because the hormone is bound to the antibody, but there is a reduction in the amount of free hormone available to target tissues.

In species that usually give birth to a single offspring (monotocous

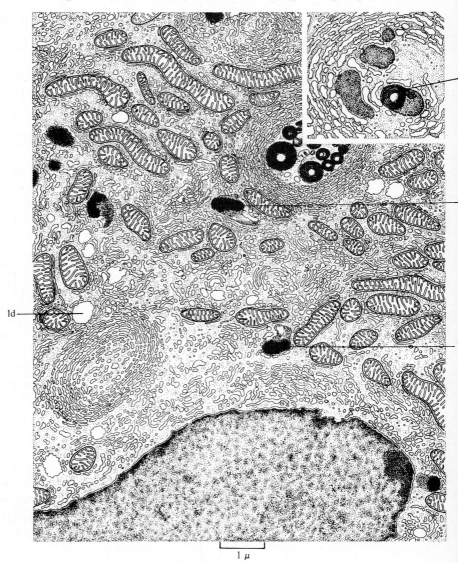

ld

1 μ

species), such as man and the cow, the number of corpora lutea usually corresponds to the number of embryos, but in the mare the corpus luteum of pregnancy is supplemented by several accessory corpora lutea, formed early in gestation either by ovulation or by luteinization of unruptured follicles. In highly polytocous species, such as rabbits and pigs, some embryonic mortality commonly occurs and the number of corpora lutea usually exceeds the number of embryos. An extreme example of such a discrepancy is found in the plains viscacha, a hystricomorph rodent from S. America related to the guinea pig, in which several hundred follicles ovulate at a time, giving rise to a mass of lutein tissue, although only one or two blastocysts normally implant and develop (see Chapter 5).

Progesterone is the most active of the naturally occurring progestogens and is produced by the ovary of most species in greater quantities than any other steroid. The major site of production is the corpus luteum, although in some animals the ovarian interstitial tissue may be a significant source. The corpus luteum of pregnancy is composed of morphologically characteristic cells, large and round in appearance and well-endowed with a pale-staining cytoplasm; in some species the lutein cells derived from the granulosa and the theca may be morphologically distinct, differing particularly in size. The ultrastructure of actively secreting luteal cells resembles that of the other steroid-secreting cells: the cytoplasm is packed with smooth endoplasmic reticulum and contains lipid droplets and an enlarged Golgi complex (Fig. 7.8). The lipid droplets probably represent intracellular stores of steroid precursors, depleted during maximum synthesis, as in pregnancy, and restored at times of reduced steroid production as during luteal regression.

We must consider as distinct processes the maintenance of the corpus luteum as a structure, its synthetic abilities, and its secretory activity in terms of progesterone and other metabolites. Fig. 7.9 illustrates three variants of the growth and function of the corpus luteum encountered in pregnancy. It may grow larger, survive longer and be more active than during the normal cycle or pseudopregnancy (rat and guinea pig); alternatively, its size and activity may not increase, although its life-span is prolonged (sheep, cow and pig). Finally, its life-span, growth and function may be indistinguishable from a corpus luteum of pseudopregnancy (ferret, dog). So far as other ovarian structures are concerned, follicular maturation and ovulation are usually suspended, at least during late pregnancy.

The maintenance of luteal function in pregnancy is largely regulated by the continued secretion of a luteotrophic hormone complex. In some species (rabbits, pigs and goats) this complex of protein hormones must be derived principally from the pituitary, since hypophysectomy always results in abortion (Table 7.1). In other animals (sheep and rats) it appears that the pituitary is only necessary at the beginning of pregnancy, since placental luteotrophins are able to compensate for its removal later on.

The early demonstration in hypophysectomized pregnant rats that injections of prolactin could maintain functional corpora lutea and prevent abortion led to the view that prolactin was the luteotrophic hormone. More recently it has become apparent that in most species the luteotrophic stimulus is made up of a hormone complex in which prolactin plays an important part. A consideration of the nature of this luteotrophic complex in six species will illustrate the diversity of the endocrine mechanisms that have developed to control luteal function during gestation, as well as the parallels underlying their modes of action.

Rabbit. The corpora lutea soon degenerate after hypophysectomy, and prolactin plus FSH and possibly low levels of LH are necessary to support luteal function. The major role of this luteotrophic complex is to stimulate the follicles to secrete oestrogens, which have a direct trophic influence on luteal cells, prolonging their life and promoting progesterone secretion. This dependence on follicular oestrogen has been neatly demonstrated by Landis Keyes and his colleagues by destroying the follicles with X-irradiation, which promptly causes luteal regression.

Rat. Prolactin and LH are the chief components of the pituitary luteotrophic complex during the first half of gestation in the rat, and they are supplemented during the second half by a prolactin-like luteotrophin secreted by the placenta. Furthermore the uterus responds to a decidualizing stimulus by producing a luteotrophic substance which ensures the longevity

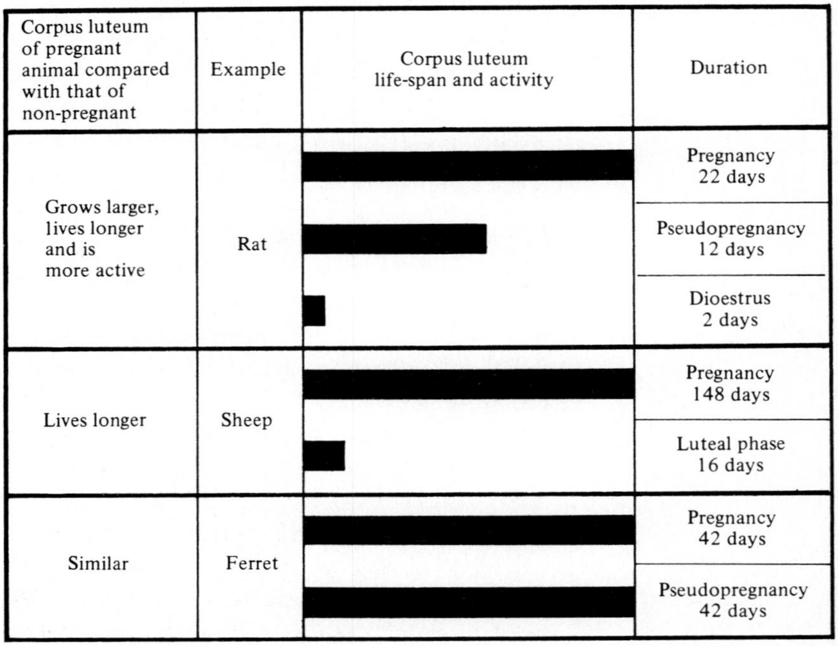

Fig. 7.9. Three variants of the growth and function of the corpus luteum in pregnancy.

Corpus luteum of pregnant animal compared with that of non-pregnant	Example	Corpus luteum life-span and activity	Duration
Grows larger, lives longer and is more active	Rat		Pregnancy 22 days
			Pseudopregnancy 12 days
			Dioestrus 2 days
Lives longer	Sheep		Pregnancy 148 days
			Luteal phase 16 days
Similar	Ferret		Pregnancy 42 days
			Pseudopregnancy 42 days

of the corpus luteum; subsequently the placenta itself may produce a chorionic gonadotrophin. Geula Gibori and JoAnne Richards have shown that the action of prolactin (and subsequently placental lactogen) is exerted through the maintenance of luteal receptors for LH and oestradiol, and that an important action of LH is to maintain synthesis of oestradiol by the corpus luteum itself. The oestradiol produced in response to LH is thought to stimulate progesterone synthesis by an effect on an as yet unidentified step in steroidogenesis from cholesterol. Thus the actions of prolactin or, later in pregnancy, of placental lactogen, result in maintained synthesis of oestradiol, which is the key to maintained progesterone secretion; the system therefore has some similarities to that in the rabbit. The importance of oestradiol is evident from experiments in which an aromatizable androgen, such as testosterone, can be used in the absence of any other replacement therapy to maintain luteal function in rats which have been hypophysectomized and hysterectomized during pregnancy; non-aromatizable androgens, such as dihydrotestosterone, are without effect.

Sheep and goat. Although these two ruminants are closely related phylogenetically, being derived from a common ancestral stock and having morphologically identical forms of placentation, they demonstrate the considerable differences that can exist between species as regards the importance of luteal progesterone secretion in pregnancy maintenance. In both animals the placenta secretes large quantities of C_{21} steroids and oestrogens, and each species produces a placental lactogen (Fig. 7.10). However, there is a striking difference in the nature of the C_{21} steroids produced by the placenta: the goat's placenta secretes large quantities of pregnanediol, derived by reduction at the 3, 5 and 20 positions from progesterone synthesized in the placenta. In the sheep's placenta these reductive activities are less pronounced, so that progesterone is the major C_{21} steroid produced. The outcome is that the corpus luteum of the sheep is no longer the most important source of progesterone, and can be dispensed with after day 50 of gestation without causing abortion, whereas the corpus luteum of the goat is essential throughout pregnancy.

The increasing rate of placental progesterone secretion in sheep during pregnancy is accompanied by a gradual decline in luteal function so that by the end of gestation the corpus luteum has usually regressed. During the first half of gestation, however, there is evidence for the production of a luteotrophic hormone which may be a placental gonadotrophin (Fig. 7.10). In the goat on the other hand, luteal function is maintained throughout gestation until shortly before parturition, and in fact it increases during the second half of pregnancy as a result of secretion of placental lactogen. As Hugh Buttle has shown, placental lactogen can replace pituitary prolactin as a component of the luteotrophic complex after mid-gestation, since, following hypophysectomy, pregnancy can be maintained by LH alone after about day 80. It should be noted that the continued dependence on pituitary LH is not conclusive evidence against the existence of a chorionic gonadotrophin in the goat, but it does indicate that any such hormone is not produced in sufficient quantities to maintain luteal function in the absence of the pituitary.

Pig. Luteal function in the pig resembles that in the goat, being indispensible for pregnancy maintenance throughout gestation. During early pregnancy the corpus luteum is dependent on LH; from about day 70 it can be maintained after hypophysectomy with prolactin. The corpus luteum of the pig produces not only progesterone but also the important peptide

Fig. 7.10. Interrelationships between hormones of the placenta, corpus luteum and pituitary in goat and sheep. Goats are dependent upon luteal progesterone for pregnancy maintenance and a negative feedback loop exists between the corpus luteum and the hypothalamus by which progesterone gradually reduces LH secretion. Luteal function is also maintained by caprine placental lactogen (cPL), and the placenta produces a pregnanediol, but little progesterone. In sheep, the placenta has partially lost the ability to reduce progesterone to pregnanediol, and as a result it secretes large amounts of progesterone. This process is not influenced by the circulating level of LH. However secretion of pituitary luteotrophin (LH) is inhibited by progesterone, and in consequence luteal function is gradually reduced in late pregnancy. Ovine placental lactogen (oPL) and possibly an ovine chorionic gonadotrophin (oGC) are produced but are not sufficiently luteotrophic to prevent luteal regression during the second half of pregnancy. Thickness of arrows denotes degree of functional importance.

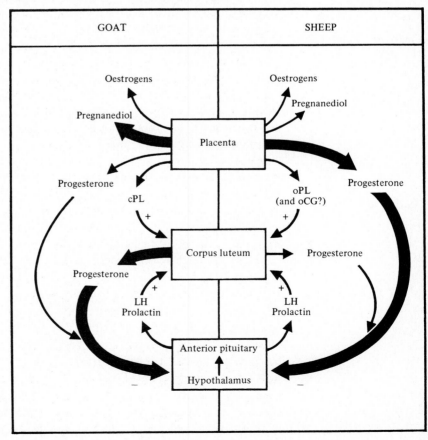

hormone, relaxin. This polypeptide (molecular weight 6000), which has a striking structural resemblance to insulin and epidermal growth factor, is primarily involved in controlling cervical dilatation at term, and is released from the corpus luteum in large quantities during the last 2 to 3 days of pregnancy. The stimulus for luteal release of relaxin is thought to be uterine $PGF_{2\alpha}$.

The importance of relaxin for the normal delivery of piglets has been demonstrated by David Sherwood in Illinois, who recorded high still-birth rates in sows in which the supply of relaxin was removed by ovariectomy. When ovariectomized animals were treated with progesterone to maintain pregnancy, and the therapy discontinued at the expected time of farrowing, piglet survival rate was only 48 per cent, but when both progesterone and relaxin were given this was raised to 98 per cent. Relaxin may also have other effects on the uterus and mammary gland; in the rat, relaxin extracted from porcine corpora lutea is a potent inhibitor of uterine contractions.

Man. The maintenance of the corpus luteum of pregnancy in women probably depends initially on the secretion of human chorionic gonado-trophin (hCG) by the trophoblast shortly after implantation, and a placental lactogen (hPL) may also be involved later in gestation. Continued luteal function is required until 6–7 weeks of gestation, as indicated by failure of pregnancy in patients undergoing ovariectomy or lutectomy before this time. The human corpus luteum is unusual in that it is capable of synthesizing large quantities of oestrogens, both in the cycle and early pregnancy. The circulating levels of a precursor of oestrogens, 17α-hydroxyprogesterone, can be used as an indicator of luteal function in early gestation as this compound is produced by the corpus luteum but not by the trophoblast.

Hormone production by the placenta

In some species the placenta may assume some of the endocrine functions of the ovary. This is best seen in animals where ovariectomy does not interfere with pregnancy (Table 7.1). In women and rhesus monkeys, the placenta produces enough progesterone to maintain pregnancy from a very early stage, so that the role of the corpus luteum in pregnancy maintenance is short-lived. In other animals, like the pig, goat and rabbit, the placenta never secretes sufficient progesterone to maintain gestation in the absence of the ovaries.

There are great species differences in placental endocrine function, and there appears to be little correlation with placental morphology. Nonetheless it seems likely that the endocrine differences between some species may be more quantitative than qualitative. For instance, it has recently been shown that although the placenta of the pig and the goat produce insufficient progesterone to maintain pregnancy after ovariectomy, they do secrete some progesterone. Furthermore most placentae also secrete oestrogens.

The placenta may also take over some of the functions of the pituitary by the production of chorionic gonadotrophins. Human chorionic gonadotrophin (hCG), which was discovered by Aschheim and Zondek in 1927, is a glycoprotein made up of two non-covalently bonded subunits (α and β). The α-subunit, with a molecular weight of 15000, shows considerable homology in its amino acid sequence with that found in LH, FSH and TSH, and the β-subunit, with a molecular weight of 23000, shows a considerable homology with the β-subunit of human, porcine, bovine and ovine LH. The individual subunits are biologically inactive, but when recombined they recover the biological and immunological activity of the parent hormone almost completely. Carbohydrate constitutes about 30 per cent of the hCG molecule and 5 of the 7 carbohydrate units are associated with the β-subunit (see Table 7.3). In many mammalian species, hCG binds to membrane receptors for LH on the cells of gonadotrophin target tissues sited in the ovary and testis. The binding of hCG and LH to the purified receptor is specific and saturable, and dose-dependent competition between labelled and unlabelled hormone has formed the basis of radioreceptor assays for gonadotrophic activity. Om Bahl has removed different carbohydrate units from the molecule using specific enzymes and shown that glycosidase-treated derivates retained most of their binding and immunological activity, although their biological activity measured *in vitro* was reduced. Removal of certain carbohydrate units, however, increased the rate at which the hormone was metabolized in the body (Fig. 7.11), and future work is being directed towards prolonging the half-life of these derivatives by chemical means. The discovery of a potent antagonist of natural hCG with a high affinity for LH receptors on the corpus luteum,

Table 7.3. *Properties of hCG and its subunits*

Property	hCG	hCG-α	hCG-β
Molecular weight*	38000	15000	23000
% Carbohydrate	33	33	33
No. of amino acid residues	239	92	147
No. of sugar residues	57–65	22–26	35–39
No. of carbohydrate units	7	2	5
Biological activity (IU/mg)†	15000	5	5
Receptor binding activity (IU/mg)‡	15000	4	2
% Immunological activity	100	1–2	10

* Determined from the amino acid sequences and the carbohydrate composition of the α- and β-subunits. † Determined by *in vivo* bioassay.
‡ Determined by radioreceptor assay.
(From O. P. Bahl, *Fedn. Proc.* **36**, 2119–27 (1977), with additional data from L. E. Reichert, F. Leidenberger and C. G. Trowbridge, *Recent Prog. Horm. Res.* **29**, 497–526 (1973).)

but of low biological activity, offers considerable promise as a way of preventing the establishment of gestation.

We have seen that hCG is important in the establishment of pregnancy in women because of its luteotrophic properties. In the human and other primates the concentration of chorionic gonadotrophin in placenta, serum and urine is high. In man the hormone is synthesized in the syncytiotrophoblast, and its production rises steeply after implantation. Bruce Hobson and Leif Wide have recently detected chorionic gonadotrophin in placental extracts from 11 ape and monkey species using bioassay and radioimmunoassay techniques. These workers claim that the structure of the CG compounds is similar to that of human CG in late pregnancy, and that the concentration of CG in most placentae at term is the same as that in the human placenta at a comparable stage of pregnancy. A typical example of the chromatographic separation of chorionic gonadotrophin and its component units from human and gorilla placenta is shown in Fig. 7.12. Similar chorionic gonadotrophins have been detected in the placentae of rats, mice and guinea pigs, and this work strongly suggests that such placental gonadotrophins will be found widely distributed among mammals. Whereas in early pregnancy the hormone seems to be primarily concerned with the maintenance of the corpus luteum, at later stages its role is not

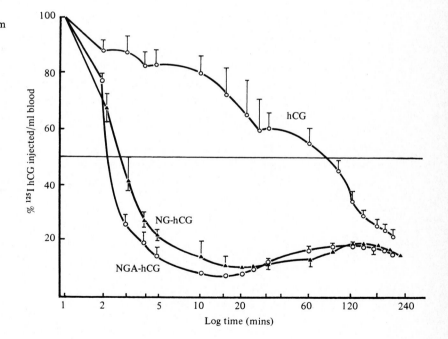

Fig. 7.11. Removal of carbohydrate residues from human chorionic gonadotrophin increases the rate at which the hormone is metabolized after injection into dioestrous female rats. The purified hormone, hCG, was treated enzymatically to remove 100% sialic acid and 60% galactose (NG-hCG), or 100% sialic acid, 60% galactose and 55% glucosamine (NGA-hCG). The hormone and its derivatives were labelled with radio-iodine (^{125}I), and blood samples were removed and counted at different times after injection. (From S. K. Batta, M. A. Rabovsky, C. P. Channing and O. P. Bahl. In *Ovarian Follicular and Corpus Luteum Function*, ed. C. P. Channing, J. M. Marsh and W. A. Sadler. Plenum Publishing Corporation; New York (1979).)

known. Claims that purified hCG suppresses lymphocyte responsiveness *in vivo* and *in vitro* have been vigorously disputed, leaving a quantitatively important hormone without a clearly understood function after the onset of placental progesterone synthesis. Qualitative tests for hCG in urine or blood are used to diagnose pregnancy in women, and hCG assays are invaluable in the clinical management of trophoblastic tumours, such as choriocarcinoma, which produce substantial amounts of this hormone.

Little is known of the factors controlling placental endocrine function. In man, production of oestrogens such as oestradiol and oestriol by the placenta is dependent upon a supply of substrates for aromatization: dehydroepiandrosterone sulphate for oestradiol synthesis is supplied by the maternal and fetal adrenals, and some of this is hydroxylated at position 16 in the fetal liver, for subsequent conversion by the placenta to oestriol (Fig. 7.13). Evan Simpson and his colleagues in Dallas have shown that an important substrate for fetal adrenal steroid synthesis is cholesterol bound to low-density lipoprotein (LDL), which is produced by the fetal liver; while this accounts for 70 per cent of cholesterol used by the fetal adrenal, the remainder is derived from the maternal circulation and the adrenal itself. The interdependence of the fetal and placental compartments in the production of oestrogens has led to the concept of

Fig. 7.12. Chromatographic properties of chorionic gonadotrophin and its α- and β-subunits purified and prepared from the placenta of (*a*) woman, and (*b*) gorilla. The proteins were mixed and chromatographed on columns of Sephadex G-200, and each fraction was assayed in hCG, hCG α-subunit and hCG β-subunit radioimmunoassays. The elution patterns obtained after chromatography are similar. (From B. M. Hobson and L. Wide, *Folia Primatol.* **35**, 51–64 (1981).)

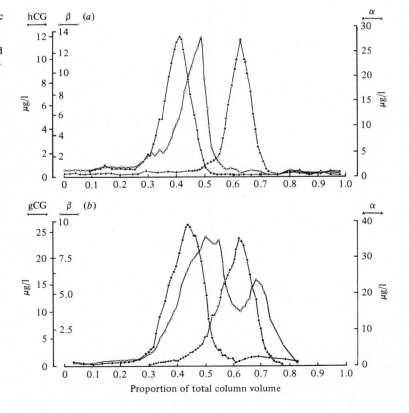

a feto-placental unit, in which both components are necessary for oestrogen production. This provides the obstetrician with a convenient method for monitoring fetal well-being *in utero*, by the measurement of oestriol in the mother's blood or urine.

The feto-placental unit seems to be a characteristic of primates, and is not generally involved in oestrogen production in other species, possibly because fetal zones in the adrenal cortex are not widely distributed phylogenetically. In the absence of control of endocrine function by supply of substrate, other mechanisms presumably apply in other animals. The mechanism leading to the onset of progesterone secretion by the sheep's placenta at about day 50 of pregnancy, or to the initiation of pregnanediol synthesis by the goat's placenta at a similar stage, are not known, though they seem to be associated with a general increase in endocrine function at this time. Secretion of these C_{21} steroids is accompanied by increased production of placental oestrogens and placental lactogen. Although the increased secretion of these hormones only becomes manifest at mid-gestation, the onset of their synthesis in the trophoblast occurs much earlier. Since there is no obvious morphological change in the placenta that would explain this sudden hormone secretion into the blood, biochemical or endocrine explanations have been sought. One of the few mechanisms known to influence placental endocrine function is the induction, by rising fetal cortisol secretion, of certain catabolic enzymes in the placentae of sheep and goats; this will be considered later in relation to the onset of parturition.

Fig. 7.13. The human feto-placental unit, showing the interdependence of the fetal and maternal compartments in the production of oestriol. Cholesterol bound to low-density lipoprotein (LDL) is supplied to the fetal adrenal cortex principally by the fetal liver (and to a lesser extent from the maternal compartment). Here it is converted to dehydroepiandrosterone sulphate which is then 16-hydroxylated by the fetal liver to form 16α-hydroxydehydro-epiandrosterone and its sulphate. This is the major substrate for placental aromatization to oestriol, which is quantitatively the most significant oestrogen in the maternal circulation.

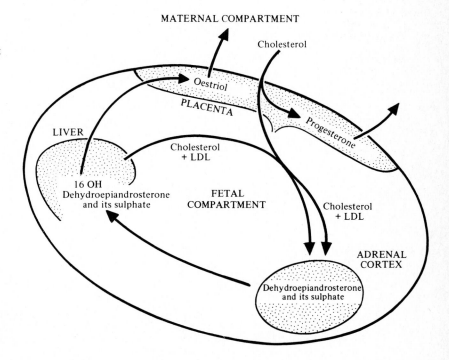

An intriguing puzzle for reproductive biologists is posed by the horse, and in particular by findings that show that the production and composition of its placental gonadotrophin, equine chorionic gonadotrophin (eCG, formerly called pregnant mare's serum gonadotrophin, PMSG), are influenced by fetal genotype. Equine chorionic gonadotrophin is produced by discrete structures on the inner surface of the uterus, the endometrial cups. In 1972 Twink Allen and his colleagues in Cambridge showed conclusively that the cups are formed by fetal rather than maternal cells, and the formation, growth and rejection of these specialized structures is of considerable biological importance.

By the end of the first month of pregnancy, the horse conceptus has grown to about the size of an orange. As the developing allantois begins to grow out, it displaces the primitive yolk sac placenta and forms the definitive allantochorionic placenta (see also Book 2, Chapter 2, Second Edition). An allantochorionic girdle develops at the interface of the allantochorion and yolk sac, and this extends around the circumference of the embryo (Fig. 7.14). Cells begin to detach from the girdle, starting on about the 36th day of gestation. They penetrate the maternal endometrium and burrow deep into the stroma, where they enlarge to form the characteristic 'decidual' cells of the endometrial cup (Fig. 7.15). On about

Fig. 7.14. Appearance of a normal horse embryo and its membranes in section and in surface view on (*a*) the 28th and (*b*) the 35th day of gestation. The whole conceptus is about the size of an orange. Cells become detached from around the chorionic girdle at this stage to invade the endometrium and form the endometrial cups.

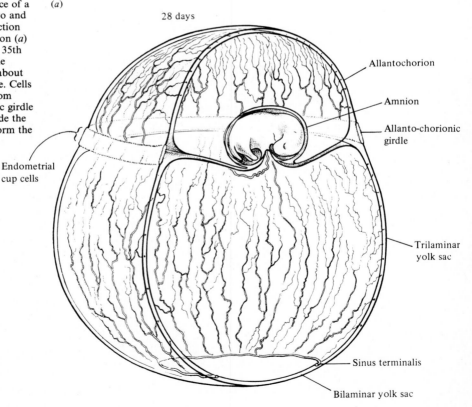

(*a*)

28 days

Allantochorion

Amnion

Allanto-chorionic girdle

Endometrial cup cells

Trilaminar yolk sac

Sinus terminalis

Bilaminar yolk sac

the 40th day, the outline of the endometrial cups becomes visible to the naked eye on the inside lining of the uterus, and eCG becomes detectable in maternal blood. If allantochorionic girdle cells are grown in tissue culture, substantial amounts of eCG are produced. By about the 60th day of gestation eCG levels reach their maximum in blood, coinciding with maximum development of the endometrial cups. After this time the endometrial cups are progressively invaded by lymphocytes (see Fig. 7.15), their production of eCG declines, and eventually they are sloughed off from the surface into the uterine lumen by a process remarkably similar to a typical host-versus-graft rejection reaction.

But it is the effect of fetal genotype on the quantity and type of eCG produced that demands attention. If a stallion is mated to a jenny donkey to produce a hinny fetus, the hybrid endometrial cups produce six to eight times the amount of eCG found in the circulation of donkeys carrying normal donkey embryos. Conversely, if a jack donkey is mated to a mare to produce a mule fetus, the amount of eCG produced is much lower than in a mare carrying a normal horse conceptus. Gonadotrophic activity in blood disappears much earlier in a mule pregnancy, and lymphocytic infiltration of endometrial cups begins much earlier, so that the whole structure is prematurely destroyed within as little as 10–15 days. Perhaps

(*b*)

35 days

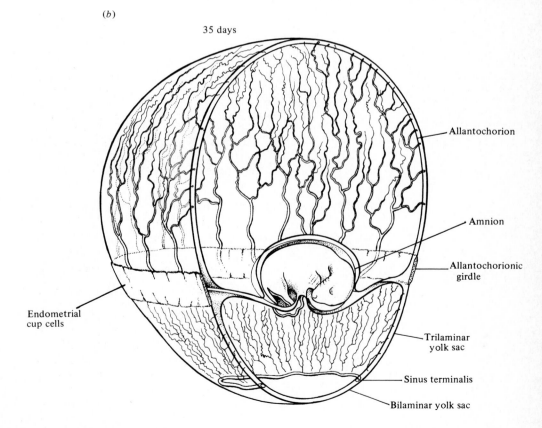

Fig. 7.15. (*a*): Appearance of the inner surface of a mare's uterus on the 60th day of gestation after removal of the embryo and its membranes. The row of endometrial cup tissue is clearly visible. (*b*): Low-power magnification of a section through an endometrial cup, showing uterine glands distended with secretion rich in eCG activity. The area between the glands is packed with 'decidual' cells. (*c*): High-power magnification of a section near the base of the cup to show the 'decidual' cells which are fetal in origin, and which produce eCG. A strong maternal immune response has been elicited as seen from the invading lymphocytes which will eventually result in the 'rejection' of the endometrial cup.

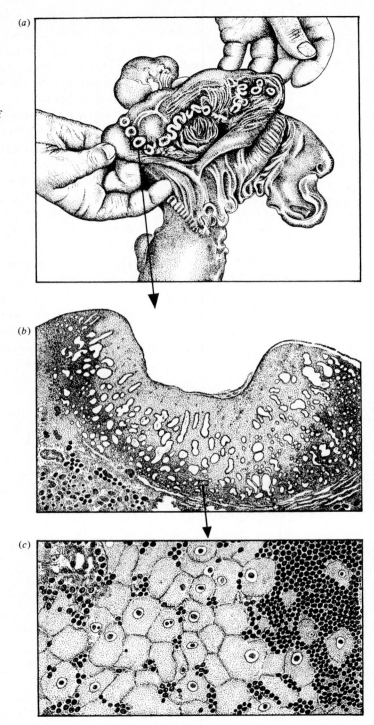

(*a*)

(*b*)

(*c*)

the mare finds donkey histocompatibility antigens more immunogenic than the donkey finds the equivalent horse antigens. Whatever the explanation, the way that the mother's immune system detects and attacks fetal cells that have invaded her endometrial stroma, whilst leaving the rest of the placenta unscathed, has wider implications for unravelling the complexities of the immunoprotection of the fetal allograft.

Regulation of eCG production by the fetal genotype is only part of the story. Circulating progestogen levels are also affected, and they too reach remarkably high values in donkeys carrying hinny fetuses (Fig. 7.16d). The eCG molecule posseses both FSH-like and LH-like activity, but when a donkey carries a hinny fetus, the resultant CG has much more FSH-like activity than has donkey CG. In the horse, and to a lesser extent the donkey, the ovary seems to be remarkably refractory to the FSH-like activity of its own CG, and although enormous amounts are produced, there is little excessive follicular stimulation of the mother's ovaries during pregnancy. Hinny pregnancies would never have occurred naturally since horses and donkeys evolved in different parts of the globe (see Book 6, Chapter 4); when they do, as a result of man's intervention, it is not surprising that this evolutionary adaptation breaks down and the ovary becomes hyperstimulated by the unusual form of CG.

So the question remains, what is the normal function of eCG in equine reproduction? Francesca Stewart and Twink Allen propose that these

Fig. 7.16. Mean plasma progesterone and serum eCG concentrations measured at weekly intervals in the blood of (a) 30 pregnant pony mares carrying normal horse conceptuses; (b) 11 mares carrying mule conceptuses; (c) 14 donkeys carrying normal donkey conceptuses; and (d) 6 donkeys carrying hinny conceptuses (note change of scale). (From Francesca Stewart and W. R. Allen. *J. Reprod. Fert.* **62**, 527–36 (1981).)

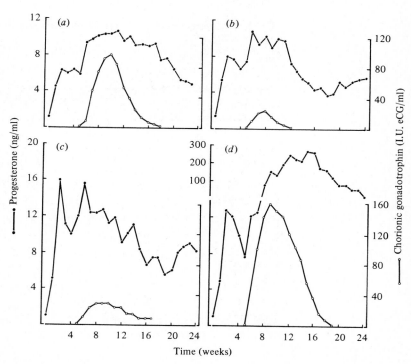

gonadotrophins do have some luteotrophic function, since they can cause ovulation and/or luteinization of accessory follicles at about day 40 of gestation. These follicles are initially stimulated by FSH of maternal pituitary origin, and after ovulation they form secondary corpora lutea whose life and secretory function are enhanced by CG and other factors. These gonadotrophic actions are LH-like, rather than FSH-like, and it is still unclear why the FSH activity is so dominant in equine CGs, especially when it is not required for successful pregnancy and may even be deleterious. Perhaps these glycoprotein hormones have some other function in helping to protect the fetal allograft from the attentions of the mother's immune system, though the ability of a mare to carry a transferred donkey embryo to full term in the absence of any detectable eCG in her circulation argues against this. But it would be premature to relegate eCG to the class of a biologically redundant molecule before we know more about it.

The cellular mechanisms by which placental lactogens are secreted in the goat and sheep provide another intriguing example of feto-maternal interactions, since the hormone is packaged in binucleate cells of the chorionic epithelium and not released until these cells have migrated across the feto-maternal junction. In this way placental lactogen is secreted preferentially into the maternal circulation, and levels in the fetal circulation are relatively low. Evidence for this migration of cells, which continues throughout gestation, has been obtained by Peter Wooding's group at Babraham using autoradiography, at the electron microscope level, after labelling nuclei of chorionic cells with tritiated thymidine. When placentae were examined at various times after injection of thymidine, the label could be traced in the nuclei of, firstly, mononucleate cells of the chorionic epithelium, then binucleate cells, and finally those in the maternal syncytium. This course of secretion, which is illustrated in Fig. 7.17, resembles that occurring in primates, where the placenta contains cytotrophoblast which differentiates into a syncytiotrophoblast. In these examples, however, the syncytium formed by fusion of fetal and maternal cells is not sloughed off and rejected by invading lymphocytes, as in the mare, but survives throughout gestation.

Hormone production by the pituitary and other endocrine organs

The maternal pituitary has an indispensable role in the secretion of hormones that are concerned with the early events of pregnancy but, as we have seen, its role is transferred in some species to the placenta in the later stages of gestation (Table 7.1). Relatively little is known about the changes that occur in the pituitary content of hormones during pregnancy, though recent work in the sheep has shown that by the end of gestation there is a very low content of pituitary LH. In animals that ovulate soon after parturition, follicular maturation must occur during late gestation, and therefore pituitary suppression is not so pronounced.

Fig. 7.17. Peter Wooding's interpretation of how granulated binucleate cells (1) could arise by cell division in the trophectoderm of the placenta of sheep and goats, and migrate across the feto-maternal junction (2–4). The binucleate cell contains granules that react with an antiserum raised against placental lactogen. After the binucleate cell has passed through the junction, the granules change from a basal (2) to an apical position (3), and their contents are then released into maternal connective tissue by exocytosis on the maternal side of the placenta (4). The binucleate cells also form a syncytium which eventually replaces the uterine epithelium. This migration of cells occurs from about the time of implantation until parturition.

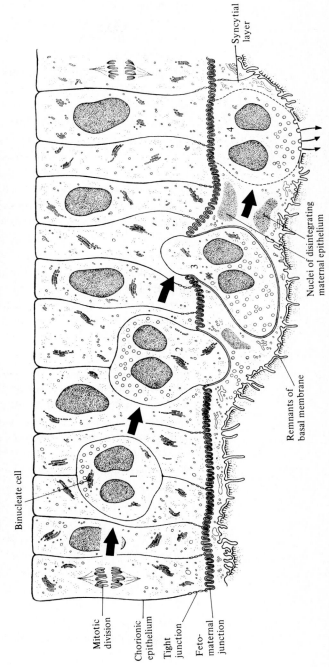

Binucleate cell

Mitotic division

Chorionic epithelium

Tight junction

Feto-maternal junction

Remnants of basal membrane

Nuclei of disintegrating maternal epithelium

Syncytial layer

Pregnancy is associated with an increase in size of the pituitary in some species. The cytology of the pituitary changes during gestation and the appearance of specialized 'pregnancy cells' is one of the most characteristic findings. The function of these cells is uncertain but may be related to increased secretion of a pituitary luteotrophic complex. The cells are believed to be derived from the chief, or 'chromophobe', cells.

The function of other maternal endocrine organs (thyroid, parathyroid, pancreas, adrenals) is related to the altered homeostasis imposed by pregnancy rather than to the fact that they have a major role in the maintenance of gestation. The adrenals may become more active and enlarge. Biochemically, they have the capacity to produce appreciable amounts of gonadal steroids (oestrogens, androgens and progestogens). The concentration of these steroids in adrenal venous blood is high under the conditions of stress induced experimentally by surgery, anaesthesia or ACTH administration, but whether they are high in the normal unstressed condition is doubtful. If gonadal hormones of adrenal origin played a significant part in pregnancy maintenance, then adrenalectomy would be expected to result in abortion; on the contrary, experiments first performed in the dog show that adrenalectomy is tolerated better when the animal is pregnant than at other times. This is probably because the secretory activities of the corpus luteum and placenta compensate in some way for the loss of corticosteroids. Similar results have been obtained in non-pregnant animals, which also survive longer after adrenalectomy if they are treated with progesterone.

Kinetics of hormone metabolism

The increased production of some hormones in gestation has been studied by the measurement of hormones in blood, or by the estimation of the daily excretion of hormones; more specifically, the estimation of hormone production rates has been carried out using tracer kinetic techniques. With the advent of the new and highly sensitive methods of radioimmuno-assay and enzyme-immunoassay, day-to-day changes in the circulating levels of hormones can be monitored throughout gestation. Because of their great lipid solubilities, oestrogens and progesterone are largely transported in blood bound to plasma proteins such as albumin, which has a very large capacity for steroid binding but a relatively low affinity. In consequence, only a small proportion of the oestrogen and progesterone measured in blood is present in a form that is free or non-protein-bound. The presence of plasma proteins in gestation with high affinities for steroid hormones (Table 7.4) affects the rate of metabolism of a steroid and its concentration in blood, and also its availability to target tissues. Therefore, we should consider the rates at which hormones are produced and metabolized in the body since this affects their role in gestation.

Under steady-state conditions the amount of a hormone produced is equal to the rate at which it is destroyed. The rate of destruction can be measured by several techniques, including one in which a tracer quantity of labelled steroid is infused at a constant rate for several hours. Such a tracer can be labelled with a radioactive or non-radioactive isotope. When the ratio of the isotopic and endogenous hormone concentration in blood is constant, the metabolic clearance rate (MCR) is given by the equation:

$$\text{MCR (ml/min)} = \frac{\text{Rate of infusion of isotopically-labelled compound } (\mu\text{Ci/min})}{\text{Blood concentration of isotopically-labelled compound } (\mu\text{Ci/ml})}.$$

Metabolic clearance rate is defined as the volume of blood that is completely and irreversibly cleared of a compound in unit time. The production rate of the hormone into the blood (mg/day) is calculated by multiplying the metabolic clearance rate by the blood concentration. An alternative way to measure hormone production is to determine the urinary excretion of a unique metabolite, such as pregnanediol or oestriol. Both methods have their particular applications, but the information derived from the urinary procedure is less direct and may be complicated if the excreted metabolite is derived from other precursors. Thus, urinary pregnanediol in women may be formed from several precursors, such as

Table 7.4. *Plasma proteins binding and transporting steroids during pregnancy*

Plasma protein	Molecular weight	Steroids bound	Comments
Albumin (human)	69 000	Androgens, cortico-steroids, oestrogens, progestogens	—
α_1-acid glycoprotein (human)	41 000	Progesterone, testosterone	—
Transcortin, cortico-steroid-binding globulin, CBG (human, monkey, guinea pig, rat and many other species)	52 000	Cortisol, cortico-sterone, progesterone	In some species concentration greatly increases in pregnancy; also increases in women on 'Pill'
Sex hormone-binding globulin, SHBG (human)	52 000	Oestradiol, testosterone	Concentration increases appreciably in second and third trimesters of pregnancy
Progesterone-bindin globulin, PBG (guinea pig)	88 000	Progesterone	Concentration increases in pregnancy

progesterone, pregnenolone or 20α-dihydroprogesterone; progesterone is quantitatively the most important of these, though only in pregnancy.

Progesterone. The concentrations of this steroid in blood during gestation differ appreciably between various species. There is a 100-fold rise over the non-pregnant levels in women and guinea pigs, a 10-fold increase in sheep and chimpanzees, and little or no rise over non-pregnant or pseudo-pregnant levels in rhesus monkey, pigs, cows, ferrets or dogs (Fig. 7.18).

There are at least two ways in which the progesterone requirements of pregnancy may be met – by either an increase in progesterone production (women), or a reduction in its rate of metabolism (guinea pigs). The sources of production are very similar in these two species; initially, the corpus luteum is the main site of secretion, but the placenta becomes increasingly important later in gestation. Yet the rate at which progesterone is metabolized differs greatly between them. In women, the MCR for plasma progesterone (see p. 179) is similar in non-pregnant and pregnant subjects (about 2500 l/day). The plasma progesterone concentrations increase gradually throughout gestation (Fig. 7.18), reflecting increas-

Fig. 7.18. The concentration of plasma progesterone in different species, pregnant (solid lines) and non-pregnant (dotted lines), and the histological type of placenta. The arrows indicate the time of parturition. Note the differences in scales.

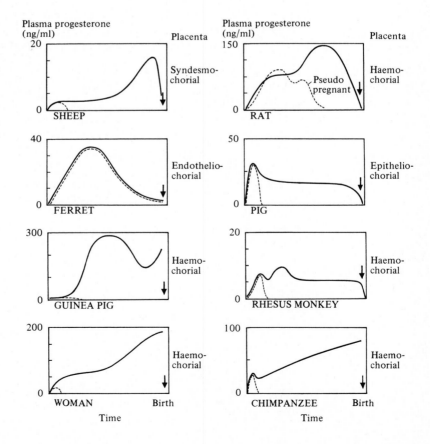

ing production which in this case is placental in origin. In the guinea pig, the progesterone concentration rises sharply after the 15th day of gestation (Fig. 7.18). The rise is accompanied by a pronounced fall in metabolic clearance rate and a raised concentration of a specific binding protein in blood which has a high affinity for progesterone (progesterone-binding globulin, PBG). This 'progesterone-conserving' mechanism has so far been demonstrated in the guinea pig, coypu, and all other hystricomorph rodents studied. It allows a large increase in the blood level of progesterone without necessarily an increased rate of secretion.

In many species progesterone is rapidly metabolized and removed from the blood stream. Investigations suggest that the steroid is distributed between two theoretical 'compartments' in the body; the rate of removal from the first compartment (possibly vascular) has a half-life of about a minute, whereas that of the second (possibly extra-vascular) about 30 minutes. In the pregnant guinea pig and coypu, the presence of progesterone-binding globulin with a high affinity for progesterone greatly decreases its rate of clearance. Progesterone production during gestation reaches a value of 250–300 mg/day in women and 1–1.5 mg/day in the guinea pig. In both species the variation in production rates between individuals is considerable. These values are appreciably greater than the daily requirements for pregnancy maintenance.

Studies of progesterone-binding globulin have compared the rates at which progesterone is bound or released by this protein – the 'on' and 'off' reactions – with its rates of reaction with a target cell receptor, as well as those of cortisol and its binding globulin (CBG) (Table 7.5). Although measured *in vitro* at a low non-physiological temperature, the results show that the rate of dissociation for PBG and CBG (k_{off}) is faster than that of a classical receptor. This accords with the receptor's role of mediating cellular responses by a prolonged retention of the hormone in the cell. The rate constant for progesterone binding by PBG is also very fast, which is consistent with a role of protection from rapid hormone metabolism.

Table 7.5. *Comparison of the kinetics of steroid binding by plasma binding proteins and a target cell receptor*

Protein	Steroid	k_{on} (M^{-1} sec^{-1})	k_{off} (sec^{-1})
PBG	Progesterone	2200×10^4	950×10^{-5}
CBG	Cortisol	20×10^4	41×10^{-5}
Chick oviduct cytosol			
– receptor subunit A	Progesterone	28×10^4	1.9×10^{-5}
– receptor subunit B	Progesterone	63×10^4	2.4×10^{-5}

(From U. Westphal, S. D. Stroupe and S.-L. Cheng, *Ann. N.Y. Acad. Sci.* **286**, 10–27 (1979).)

Oestrogens. In most species there is an increasing level of oestrogens in blood and urine throughout pregnancy (Fig. 7.19). In women this is partly attributable to the greatly elevated production of oestriol, derived mainly from the feto-placental unit (Fig. 7.13). The total quantity of oestriol excreted in urine can be up to 50 mg/day, which is about ten times more than that of oestrone and oestradiol-17β. Results available from isotope studies for the production rates of oestrogens in pregnancy indicate values of up to 30 mg/day for oestradiol-17β and oestrone. A large proportion of the oestrogens in the circulation is conjugated; in contrast, progesterone is metabolized to more polar compounds before it is excreted in a conjugated form.

A special case is that of the mare (Fig. 7.20). In the second half of gestation, when progesterone is only just detectable in the circulation and the endometrial cups have ceased to secrete eCG, there is a conspicuous increase in the oestrogens in maternal urine, in particular oestrone and the ring-B unsaturated oestrogens equilin, equilenin. These oestrogens are probably produced by a feto-placental system, in this case involving the fetal gonads rather than the adrenals, since they appear in high concentrations in maternal urine at a time when the fetal gonads undergo a conspicuous hypertrophy due to the rapid development of interstitial tissue. The horse is a good example of a species in which pregnancy is maintained by the local action of placental progesterone on the uterus, since circulating levels are extremely low though placental concentrations are notably high.

As with progesterone, oestrogens are rapidly metabolized in pregnancy both by the liver and by extra-hepatic tissues in several species. This can

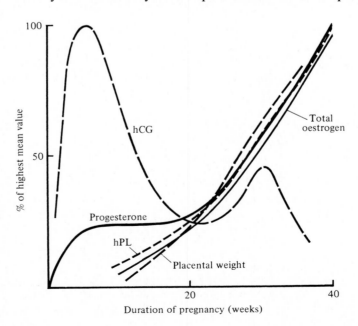

Fig. 7.19. Hormone concentrations in plasma during gestation in the human female, and their relation to placental weight. Typical values of the highest mean concentrations reported: progesterone, 200 ng/ml; total oestrogens, 150 ng/ml (individual oestrogens mainly present in a conjugated form: oestrone, 40 ng/ml; oestradiol-17β, 10 ng/ml; oestriol, 100 ng/ml); hCG, 50 i.u./ml; human placental lactogen (hPL) 10 μg/ml; placental weight, 700 g.

be deduced from measurements of metabolic clearance rate, which may often exceed hepatic blood flow. However, oestradiol-17β is concentrated in target organs such as the uterus and mammary gland by specific tissue receptor proteins (see Book 7, Chapter 6, First Edition). Progesterone is also retained by a similar mechanism, even though a large proportion is rapidly metabolized by the target organ.

Cortisol and aldosterone. The secretory activity of the adrenals appears to be modified by the changing conditions of pregnancy. In women, the concentration of cortisol in blood reaches its maximum values in the last trimester of gestation, as does the secretion rate of aldosterone. The reasons for these increases, however, are quite different.

The increased blood levels of cortisol in pregnant women are related to a reduction in the rate of cortisol metabolism, since in fact the rate of cortisol secretion actually decreases. The change in metabolism is largely attributable to a raised plasma concentration of corticosteroid-binding globulin, a protein produced by the liver that possesses a high affinity for cortisol and corticosterone, as well as for progesterone. In the non-pregnant subject, the liver is the main site of corticosteroid metabolism, whereas in pregnancy, the increased levels of corticosteroid-binding globulin result in a decreased liver metabolism and a reduction in the metabolic clearance rate of the steroid. The level of the globulin appears to be related to the high concentrations of oestrogens found in gestation, since non-pregnant patients treated with oestrogens, as in the combined oral contraceptive pill, also show an increase in this protein. Other species differ in the kinetics of corticosteroid metabolism in pregnancy, and in the rat the clearance rate

Fig. 7.20. Relative changes in hormone concentrations during gestation in the horse. Typical values of the highest mean concentrations reported: progesterone, 15 ng/ml; total oestrogens, 1 ng/ml (individual oestrogens are mainly oestrone, oestradiol-17β, equilin, equilenin and their conjugated forms); and eCG, 100 i.u./ml.

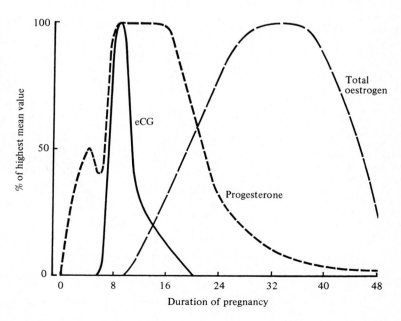

of corticosterone does not change, but its secretion rate increases near term due to hyperactivity of the fetal and maternal adrenals.

Changes in maternal adrenal function are probably not directly related to the maintenance of pregnancy. We have already noted that some mammalian species survive the stress of total adrenalectomy more successfully when pregnant or pseudopregnant than when non-pregnant. Furthermore, women with Addison's disease or those who have been bilaterally adrenalectomized can still have normal pregnancies. The observed increase in aldosterone secretion in normal gestation may be due to an inhibition by progesterone of the action of mineralocorticoid on the absorptive mechanism of the renal tubule. An increased aldosterone secretion rate could be a compensatory response to the salt-losing action of high levels of progesterone in pregnancy.

Protein hormones. In women the secretion of hormones by the anterior pituitary is greatly influenced by hormones from the placenta and ovaries. Prolactin secretion increases while FSH and LH secretion decreases as a result of the feedback effects of rising steroid levels from the feto-placental unit. Plasma prolactin levels frequently exceed those found either during lactation or in patients with reproductive disorders and hypersecretion of prolactin. A fall in FSH and LH secretion also occurs in sheep during pregnancy, and this can be mimicked by treatment with oestradiol and progesterone; this also enhances pituitary responsiveness to a standard dose of thyrotrophin releasing hormone (causing prolactin release) and decreases the response to GnRH. The role of prolactin in pregnancy is obscure but in addition to trophic effects on mammary tissue it probably maintains fetal osmolality and maternal calcium balance in the face of changing requirements. In the latter case this is achieved by enhancement of renal hydroxylation of 25-hydroxy-cholecalciferol to the more potent 1,25-dihydroxylated derivative. Clinically, the increase in pituitary size that occurs during pregnancy may produce disturbing signs reminiscent of the effects of prolactin-secreting pituitary tumours, but these usually undergo remission after delivery without treatment. Growth hormone secretion is inhibited during pregnancy, and its release in response to stimuli such as insulin-induced hypoglycaemia is impaired. Plasma ACTH concentrations are slightly increased, possibly by the production of a similar protein in the placenta. The posterior pituitary hormones are discussed by Dennis Lincoln in Chapter 2.

Pituitary protein hormones have a much greater clearance rate and a shorter half-life than placental protein hormones. Values for FSH, LH, GH and prolactin indicate that the half-life ranges from 5–30 minutes. Placental hormones such as the glycoproteins hCG and eCG are apparently produced in high concentrations and have long half-lives. Assays of hCG in urine give a production rate of up to 100 000 i.u./24 h, and eCG, a gonadotrophin not found in urine, is produced at a rate of 200 000 i.u./24 h; their half-lives are measured (respectively) in hours (up to about 40 h) and days (up to 6 days). During pregnancy the polypeptide human placental lactogen (hPL) increases 18-fold over basal prolactin levels in maternal blood, and in clinical practice the levels have been used to predict perinatal mortality, which occurs with increased frequency in mothers with low hPL values in late pregnancy.

Other placental proteins have been discovered, which are present in increasing concentration in the maternal circulation as pregnancy advances. Whether they are hormones is uncertain, but they have been implicated in immunosuppression (SP_1), inhibition of fibrinolysis (pregnancy-associated plasma protein A, PAPP-A) and the proteolytic activity of trypsin (PP5). Some of these proteins behave like carcino-embryonic antigen, and they may eventually prove useful as tumour markers.

Clearly, so far as the maintenance of gestation in women is concerned, there are complex interactions between hormones and their plasma transport proteins, brought about by the changing endocrine conditions of pregnancy. In most species the production of oestrogens, progesterone and high-affinity binding proteins all increase. It may be significant that the species in which an appreciable rise in corticosteroid-binding globulin and progesterone-binding proteins has been found in pregnancy – such as man and the guinea pig – have a haemochorial placenta, where maternal blood bathes fetal tissues, and they all show plasma progesterone levels that are much greater than those in the non-pregnant condition (Fig. 7.18).

Pregnancy diagnosis

Pregnancy may be diagnosed very early by the measurement of hormone levels; immunological tests for hCG are routinely used for pregnancy diagnosis in women, and in cows the radioimmunoassay of progesterone in milk is used as an early and non-invasive method of diagnosis. If required, the results in cows can be confirmed somewhat later in gestation by the measurement of oestrone sulphate in milk (Table 7.6).

Effects of hormones on the mother

Growth of the uterus

Though progesterone is frequently referred to as the 'hormone of pregnancy' it has also been called 'the useless hormone' because it rarely acts alone. Synergistic interactions are usually encountered when the concentrations of progesterone and oestrogens differ by about three orders of magnitude in favour of progesterone. As the difference in concentrations decreases, their actions become increasingly antagonistic. The histological changes that are observed in target organs during gestation are the result of interactions between oestrogens and progesterone. In the pregnant rabbit there is extensive proliferation of the endometrium due to the 'priming' effect of oestrogens and the synergistic action of progesterone on the luminal and glandular epithelium. This proliferation prepares the endometrium for the reception of the embryo and is associated with the events of implantation. Endometrial proliferation is a prominent feature of the progestational state in the woman, Old World primates, and the cat, ferret and rabbit, but in other species the degree of proliferation is much less pronounced (rat, mouse, guinea pig, sheep and cow). The synergistic interactions between oestrogen and progesterone cause distinctive changes in the histological appearance of other target organs during pregnancy, notably the mucification of the vagina and the growth of alveoli and ducts in the mammary glands.

The growth of the uterus during gestation represents principally an enlargement in the thickness of the myometrium. This results from growth

Table 7.6. *Typical clinical methods of pregnancy diagnosis in different species*

Species	Test	Source of material	Substance tested for, or response	Week after ovulation from which test is applicable
Woman	Immunological	Urine or plasma	hCG + LH	3rd
		Urine or plasma	β-subunit of hCG	2nd
Cow	Rectal palpation	—	Fetus	6th
	Radioimmunoassay	Milk	Progesterone	3rd week only
	Radioimmunoassay	Milk	Oestrone sulphate	15th
Ewe	Radiography	—	Fetus	8th
	Ultrasonics	—	Fetus	9th
	Steroid assay	Urine	Oestrone	3rd week only
	Radioimmunoassay	Plasma	Progesterone	16–18 days only
Sow	Rectal palpation	—	Enlargement of middle uterine artery	4th
	Radiography	—	Fetus	12th
	Ultrasonics	—	Fetus	8th
	Histological test	Vaginal mucosa	Cell height	4th
	Radioimmunoassay	Plasma	Progesterone	15–21 days only
	Radioimmunoassay	Plasma or urine	Oestrone sulphate	3rd–4th week only
Goat	Radioimmunoassay	Milk	Progesterone	3rd week only
	Radioimmunoassay	Milk	Oestrone sulphate	8th
Mare	Ultrasonics	—	Fetus	3rd
	Rectal palpation	—	Fetus	3rd
	Immunological	Plasma	eCG (PMSG)	6th–14th
	Steroid assay	Urine	Oestrogens	20th

in size of existing smooth muscle cells, rather than from an increase in cell numbers. So far as the uterus is concerned, we have a good understanding of how oestrogens act to influence its growth and composition. If ovariectomized rabbits are injected with oestradiol-17β, protein synthesis is stimulated and the RNA:DNA ratio of the endometrium increases, as in normal pregnancy. If progesterone is injected, there is no effect on uterine growth, whereas treatment with a suitable combination of both hormones may promote even larger increases than are found with oestrogen alone, due to receptor-mediated events that we have discussed in Chapter 6 of Book 7 (First Edition). This influence of oestrogens on the growth and weight of the uterus is a rapid and direct effect; within 30 minutes of injection into rats, uterine hyperaemia occurs, and within 48 hours, dry weight, protein synthesis and glycogen deposition all increase progressively.

Uterine growth in pregnancy cannot be entirely accounted for by the effects of oestrogen and progesterone. The contractile protein actomyosin was first demonstrated in human uterine smooth muscle 30 years ago, and its synthesis in the myometrium is regulated to a large extent by oestrogen, though during gestation the 'stretch' induced by the growing fetus and its associated membranes also plays an important part.

Uterine blood flow increases with the growth of the uterus during gestation, but when corrections are made for the increased weight of the uterus and its contents, the relative flow apparently decreases in mid-pregnancy to fairly stable levels, which are maintained until parturition (as in the sheep and goat). Uterine blood flow is probably influenced by hormonal factors as well as by the demands of the growing fetus on maternal nutritive supplies, and in sheep the uterine vasculature is highly

Fig. 7.21. Uterine blood flow in the ovariectomized sheep is increased after a single injection of oestradiol-17β, with a lag period of about 30 min (R, right uterine horn). Adenosine infusion increases blood flow almost immediately on the contralateral side (L, left uterine horn). (R. Resnik, A. P. Killam, M. D. Barton, F. C. Battaglia, E. L. Makowski and G. Meschia. *Am. J. Obstet. Gynec.* **125**, 201–6 (1976).)

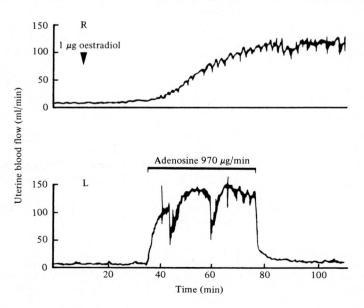

sensitive to oestrogens, which cause a striking decrease in vascular resistance and increase in blood flow. A single injection of oestradiol causes a marked rise in uterine blood flow after a lag phase of about 30 min (Fig. 7.21). Comparable increases in flow can be produced almost immediately with certain prostaglandins, nucleotides, nucleosides and other vasoactive stimuli, so that the lag in response after an oestradiol injection is probably related to a receptor-mediated synthesis of endogenous compounds that produce vasodilatation.

Control of myometrial activity

Progesterone has multiple effects on the mother in relation to the maintenance of gestation. It delays ovulation by the suppression of pituitary LH secretion; it acts on the endometrium to prepare for the reception of the embryo; and it renders the contractile myometrium quiescent so that implantation proceeds normally and the expulsion of embryos is prevented (Fig. 7.22). The late Arpad Csapo called the latter property the 'progesterone block' of myometrial activity.

Experiments in rabbits first showed that when the uterus is dominated by the influence of progesterone, as in pregnancy or when the steroid is applied locally to the uterus *in vivo* or *in vitro*, coordinated myometrial contraction is reduced. Moreover, the stimulatory effect of oxytocin on uterine contractility is abolished, so that oxytocic agents will often induce effective labour only in late pregnancy when progesterone is falling. Yet

Fig. 7.22. Spontaneous myometrial contractions in (*a*) a non-pregnant sheep, 15 days after ovulation when the corpus luteum is regressing, and (*b*) a pregnant sheep at 15 days of gestation, when the corpus luteum is maintained. Myometrial activity was measured by a fluid-filled open-ended cannula placed in the uterine lumen. (From I. R. Fleet and R. B. Heap. *J. Reprod. Fert.* **65**, 195–203 (1982).)

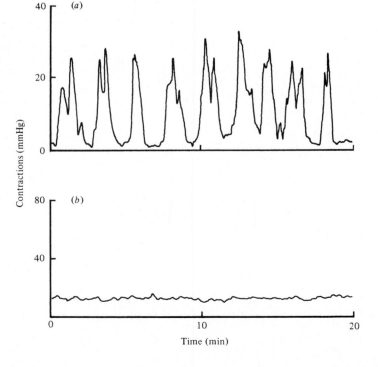

another investigation suggested that progesterone acted not only locally to reduce coordinated myometrial contractility and electrical activity, but also inhibited the hypothalamic release of oxytocin.

The 'progesterone block' hypothesis may not apply equally in all species, since although effects of progesterone on myometrial contractility have been reported in the rat, sheep and man, they are not always the same as those found in the rabbit. Arpad Csapo was intrigued by an experiment of nature where a woman had delivered twins six weeks apart. He interpreted this as further evidence for a local inhibitory effect of placental

Fig. 7.23. Relaxin inhibits uterine activity in the guinea pig. The graphs show frequency of intra-uterine pressure cycles before, during and after (*a*) cross-circulation of blood from pregnant into non-pregnant guinea pigs, and (*b*) intravenous infusion of relaxin (1 mg/h) to non-pregnant guinea pigs. Control animals received either (*a*) blood from a non-pregnant or male guinea pig, or (*b*) infusion of gelatin. The period of cross-circulation or relaxin infusion is indicated by the shaded bar. (From D. G. Porter. *Biol. Reprod.* **7**, 458–64 (1972).)

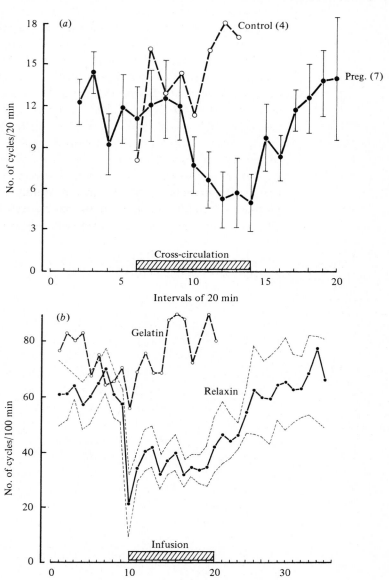

progesterone on myometrial contractility. Clinical attempts to postpone premature labour with large doses of progesterone have been relatively unsuccessful, and this finding, together with other observations, has led to scepticism that the progesterone block hypothesis applies in women or other primates. In the guinea pig, progesterone has little effect on myometrial activity, even when it is applied locally in very high doses. In contrast, David Porter has shown that relaxin is a potent inhibitor of uterine activity (Fig. 7.23) and it may therefore be an important hormone in the maintenance of pregnancy in this and other species. Oestrogen treatment, and to a greater extent combined oestrogen and progesterone treatment, of ovariectomized female guinea pigs leads to the accumulation of relaxin in the uterus. Evidence for a uterine origin for relaxin as an inhibitor of myometrial activity in women is not yet available, and studies with pure human relaxin will be watched with interest.

Nevertheless, the role of progesterone in reducing myometrial activity in the rabbit, sheep and pig is convincing, and widely differing theories have been put forward to explain how this happens. Csapo proposed that progesterone 'hyperpolarizes' the myometrial cell, reducing its excitability and the conduction of impulses between cells. So a decline in progesterone, as at term in some animals, would aid myometrial excitation. Trains of action potentials appear to originate in a group of cells called pacemakers, and with declining progesterone levels and an increased number of gap junctions between smooth muscle cells, impulse propagation and synchronized contraction develops. Changes in the ionic permeability of the myometrial cell induced by progesterone, and in the availability of intracellular calcium, seem to underlie the inhibition of activity.

Metabolic changes

Pregnancy is an anabolic state and is associated with an increased metabolic activity in the maternal organism. The characteristic gain in weight is due partly to the growth of the uterus and the products of conception, and partly to increases in body fluids, maternal stores of lipid and the growth of certain organs such as the liver and mammary glands (Fig. 7.24). In gestation the increased body retention of water, protein and fat is hormonally regulated. Thus, maternal weight gain in pregnancy is not simply an accumulation of excess reserves to protect the mother against the ever-increasing demands of the fetus and the anticipated needs of the neonate, but results also from the physiological adjustments of pregnancy controlled by the changing hormonal conditions.

Progesterone is believed to play an important part in maternal weight gain during gestation, but the mechanism of its effect is obscure. The mouse and rat have been studied in greatest detail, but even in these two species there are different explanations of the way progesterone can influence body weight. In the mouse, progesterone induces water retention, stimulates appetite and food intake, and has a protein anabolic effect. In the rat, weight gain is only observed if ovarian function is excluded by ovariectomy. Yet another effect is observed in the human female, where progesterone is catabolic and may promote increased salt excretion, though this is compensated for by increased concentrations of potential salt-retaining hormones such as oestrogens, placental lactogen and prolactin. Recent

Fig. 7.24. Cumulative weight gain during pregnancy. Weight gained is due partly to the growth of the products of conception, and partly to changes in the uterus and other maternal compartments (From F. E. Hytten. In *Clinical Physiology in Obstetrics*, ed. F. Hytten and G. Chamberlain. Blackwell; Oxford (1980).)

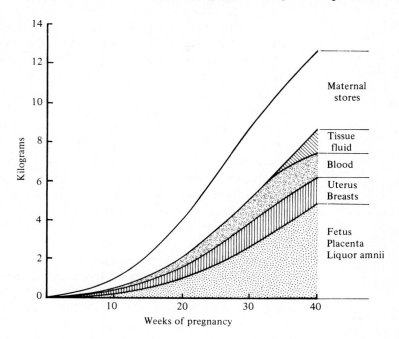

Fig. 7.25. Summary of actions of human placental lactogen on carbohydrate metabolism. The fetal hypothalamus and pituitary may be responsible for release of placental lactogen; secretion may also be affected by blood glucose level in the placenta, which is determined by the influence of the lactogen on maternal carbohydrate metabolism. Placental blood glucose levels may also control fetal carbohydrate metabolism and growth. FFA, free fatty acids. (From M. M. Grumbach, S. L. Kaplan, J. J. Sciarra and I. M. Burr. *Annals N.Y. Acad. Sci.* **148**, 501–31 (1980); D. G. Porter. *Placenta* **1**, 259–74 (1980).)

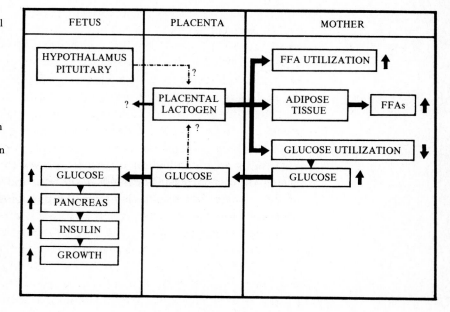

reports suggest that some women taking steroidal contraceptives gain weight, so it is obvious that the metabolic effects of progesterone and oestrogens are complex; they may even differ according to the stage of gestation.

Normal pregnancy alters glucose homeostasis, probably by a multiplicity of effects of different hormones. Raised levels of plasma cortisol in women are thought to stimulate the degradation of insulin by the liver and placenta. Human placental lactogen (hPL) possesses glucose-sparing, free fatty acid mobilizing activity, and may render more glucose available to the fetus (Fig. 7.25). The increase in the concentration of hPL in peripheral blood is similar to the pattern of change in placental weight. Ovine placental lactogen has been shown to stimulate the production of growth-promoting agents such as the somatomedins in hypophysectomized rats, a finding that implies that placental lactogen may influence fetal growth. The altered hormonal status of pregnancy causes increased levels of plasma glucose, and in women this change, together with increased insulin levels in the circulation and the tissue resistance to insulin, may produce what has been termed 'chemical diabetes of pregnancy'. The condition is usually asymptomatic and does not require insulin therapy.

The onset of parturition

If one of the functions of progesterone in pregnancy is to block myometrial activity, the increased contractility found at parturition should be associated with a removal of this inhibition, and this is discussed in detail in Book 2, Chapter 4 (Second Edition). As indicated earlier, the way that progesterone dominance is removed provides an important example of how placental endocrine function, at least in the sheep and goat, is influenced by the rising secretion of fetal cortisol, one of the harbingers of parturition. In the sheep, fetal cortisol stimulates the enzymes 17α hydroxylase and C-17, 20-lyase in the placenta, so leading to decreased secretion of progesterone by its increased metabolism to oestrogen. The resulting withdrawal of progesterone and increased circulating concentrations of oestrogens leads to production of prostaglandins by the uterus, and the onset of uterine contractions. In the goat at term, similar placental changes in response to fetal cortisol result in increased production of oestrogens (accompanied by a fall in circulating pregnanediol) and this causes luteal regression as a result of uterine production of prostaglandins.

Some observations do not support the view that parturition is initiated by a sudden fall in plasma progesterone levels in all species; for example, in women there is no such fall and progesterone injections cannot prevent labour. Other factors, such as changes in the binding of progesterone in the placenta, endometrium and myometrium, may also be involved. The activity of the fetal adrenal glands may induce prostaglandin release and trigger parturition in a number of species. The rise in prostaglandin levels in amniotic fluid and peripheral blood during the early stages of labour offers a clue that prostaglandins may be involved in the onset of labour in women – a finding supported by studies showing that women who take large doses of aspirin (a prostaglandin synthetase inhibitor) during pregnancy have a slightly longer gestation period compared with controls.

At, or just before, the time of parturition myometrial activity becomes synchronized and uterine contractions increase in both frequency and amplitude (see Book 2, Chapter 4, Second Edition). With the passage of the fetal head into the birth canal and the reflex release of oxytocin from the pituitary, this process is further stimulated, and leads to the expulsive stage of labour. The increase in myometrial activity in late pregnancy corresponds to the time when the concentration of oestrogens in blood reaches its highest levels (rat, guinea pig, sheep, goat and woman). The sheep shows the most striking rise in the levels of total unconjugated oestrogens about 48 hours before parturition, though smaller increases have been found in several species. This is also the time when spontaneous uterine activity and reactivity to oxytocin show a dramatic increase. Oestrogens administered in late pregnancy will also cause uterine contractions, and the removal of the 'progesterone block' of myometrial activity towards the end of gestation and raised oestrogen levels may both be components of the hormonal changes leading to normal delivery in sheep.

The administration of oestrogens during gestation causes abortion in several species, including the rat, mouse, rabbit, cat and cow. The abortifacient effect depends on the time of pregnancy when the treatment is started. Late pregnancy is much more difficult to interrupt in this way, presumably because of the greater concentrations required to overcome progesterone dominance. However, oestrogens probably do not have a similar importance in the termination of pregnancy in all species since in some they appear to be without effect on the duration of gestation, and in others they may even prolong it slightly.

In summary, progesterone plays a dominant role in the maintenance of pregnancy and in some species its withdrawal controls the onset of parturition. Other endocrine changes of varying importance for the initiation of labour include the activation of the fetal adreno-hypophysial axis and increased secretion of fetal corticosteroids; a rise in oestrogen and prostaglandin secretion; the relaxation of the pubic symphysis promoted by relaxin; and, during the second stage of labour, the release of oxytocin from the pituitary.

Pregnancy is essentially a partnership, at least from the hormonal point of view. In almost all mammals the adoption of viviparity involves an interplay between fetal and maternal endocrine systems for both the maintenance and termination of gestation. It is true that the mother supplies all the needs of the fetus, but from the moment of conception onwards the new individual plays an active part in safeguarding its own future. The conceptus may signal to the mother its arrival in the uterus;

produce the hormones that ensure its safe lodging; or furnish a stimulus that not only triggers birth, but in the end guarantees a source of food for its early days of life in the outside world by initiating a milk supply from the mother. It is in the regulation of these widely differing phenomena that hormones, acting as chemical messengers, play such a prominent role in the economy of gestation.

Suggested further reading

Human chorionic gonadotropin, its receptor and mechanism of action.
O. P. Bahl. *Federation Proceedings*, **36**, 2119–27 (1977).

Control of placental endocrine function: role of enzyme activation in the onset of labour. A. P. F. Flint and A. P. Ricketts. *Journal of Steroid Biochemistry*, **11**, 493–500 (1979).

The new placental proteins. A. Klopper. *Placenta*, **1**, 77–89 (1980).

Endocrine control of parturition. G. D. Thorburn and J. R. G. Challis. *Physiological Reviews*, **59**, 863–918 (1979).

Maternal Recognition of Pregnancy. Ciba Foundation Symposium 64 (new series). Excerpta Medica; Amsterdam (1979).

Placenta, structure and function (Symposium). *Journal of Reproduction and Fertility*, Supplement **31** (1982).

Conception in the Human Female. R. G. Edwards. Academic Press; New York (1980).

Hormonal maintenance of pregnancy. R. B. Heap, J. S. Perry and J. R. G. Challis. In: *Endocrinology, Handbook of Physiology*. American Physiological Society; Washington (1972).

Clinical Physiology in Obstetrics. Ed. F. Hytten and G. Chamberlain. Blackwell; Oxford (1980).

Relaxin: old hormone, new prospect. D. G. Porter. In: *Oxford Reviews of Reproductive Biology*, vol. 1, pp. 1–57. Ed. C. A. Finn. Clarendon Press; Oxford (1979).

Steroid–Protein Interactions. U. Westphal. Springer-Verlag; Berlin (1971).

8

Lactation

ALFRED T. COWIE

As the name implies, the mammary gland is the distinguishing feature of mammals. The mammary gland is part of the reproductive apparatus, and lactation is the final phase of reproduction. In most mammals it is an essential phase, and failure to lactate, like failure to ovulate, means failure to reproduce.

Fossil evidence suggests that mammals arose from certain therapsid reptiles some 200 million years ago; unfortunately the soft tissues are usually not preserved in fossils so we do not know when mammary glands first appeared during the course of evolution. Lactation bestowed considerable advantages on the mammalian mother over her viviparous reptilian counterpart, since it ensured an ideal diet for the young after birth; reptiles on the other hand might have had to migrate to special areas where the young could find the type of food that they required. It has even been suggested that the decline of large reptiles at the end of the Mesozoic may have been due to climatic and floral changes that reduced the type of food required by their young. Lactation also permitted birth of an animal with an immature skull and jaws, since teeth are not required for suckling; by the time weaning occurred, and teeth became essential for feeding, the jaws could be sufficiently developed to accommodate a nearly complete set of opposable teeth.

The stage of development of the mammalian offspring at birth varies greatly. In the monotremes (the platypus and echidna) which have retained the reptilian practice of laying eggs, the young that hatches from the egg is less than 2 cm long; it is a tiny fetus-like creature with fore-limbs that are well formed but with hind-limbs merely at the bud stage (Fig. 8.1*a*); of its sense organs probably only the olfactory system is functional. After hatching the 'fetal' monotreme is entirely dependent for its further existence on its ability to suck and lick its mother's milk which is ejected from the mammary glands onto a specialized area of skin, the areola, the monotremes having no nipples.

Although born and not hatched, the young marsupial also enters the world in a 'fetal' state. At birth, only the fore-limbs are well-developed (Fig. 8.1*b*) and with these this newly born mite, weighing only 5 milligrams in some species, mountaineers unaided from the birth canal to the pouch and thence to a teat to which it becomes very firmly attached. When it is sufficiently grown, it lets go of the teat periodically, and

Fig. 8.1. (*a*) The newly born echidna (*Tachyglossus aculeatus*, spiny anteater); inset shows appearance of the adult. (*b*) The newly born red kangaroo (*Megaleia rufa*) attached to a teat. Note in both species the well-developed fore-limbs, and that while the eyes are still covered with skin the nostrils are open. (From M. Griffiths. *The Biology of the Monotremes*. Academic Press, New York (1978).)

(*a*)

(*b*)

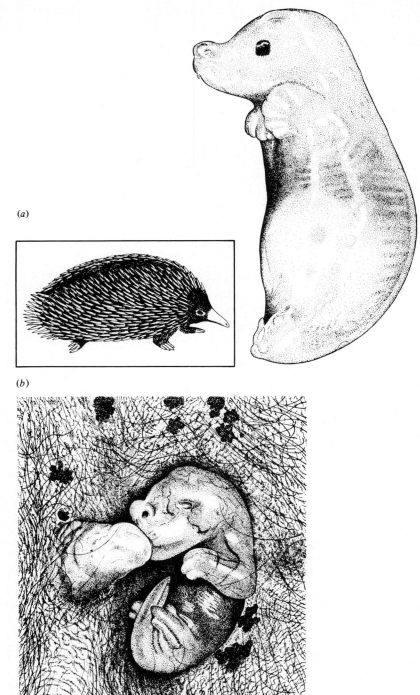

eventually makes excursions from the pouch, although returning to suckle regularly for some considerable time.

The offspring of eutherian mammals, such as mouse and man, become dependent on the mammary gland for sustenance at a much later stage of development than the young of monotremes or marsupials; the chorio-allantoic placenta has replaced the mammary gland as a source of nourishment in the early stages of life. In the guinea pig, the young are so well developed at birth that they can survive without any mother's milk if other suitable foods are available, thus making them somewhat akin to the reptiles. This, however, is exceptional, and most eutherian young are very immature at birth, and hence totally dependent on their mother's milk for a varying time thereafter.

Milk

There are large differences between species in the concentrations of the gross constituents of milk (milk fat, lactose, protein and water): milk fat may range from a trace to 500 g per litre, lactose from a trace to 100 g per litre, and protein from 10–200 g per litre. The milk fat, whose melting point must be lower than body temperature, is dispersed in the form of droplets protected from coalescing by a membrane consisting of phospholipid–protein complexes. Milk fat is a mixture of lipids – triglycerides, diglycerides, monoglycerides, free fatty acids, phospholipids and sterols; it contains a bewilderingly large array of fatty acids which also differ greatly between species. The mechanisms for controlling chain length and the extent of saturation must be highly developed and specific. Fucosyllactose and difucosyllactose are respectively the chief carbohydrates in the milks of the echidna and platypus, galactose is the chief carbohydrate of marsupial milks, and lactose of eutherian milks; however there are also lesser quantities of many other carbohydrates (such as glucose, galactose, fucose, N-acetylglucosamine) whose proportions differ widely between species. It perhaps needs to be emphasized that while the overall concentrations of the gross components of the milks of two species may be similar, the individual constitutents may differ widely. Moreover, as more refined analytical techniques become available more subtle variations are being revealed and new compounds detected.

Individuals within a species may exhibit genetic differences in milk composition; the breeds of cattle are a good example of this, since they differ widely in their butterfat content. The composition of an individual's milk also changes with the nature of the diet and with the stage of lactation. A striking example of the latter is the peri-partum change from colostrum to milk. Colostrum is the name given to the secretion of the mammary gland for several days around the time of parturition. In general colostrum has higher concentrations of protein, sodium and chloride and lower lactose and potassium than normal milk. The change from colostrum to milk occurs in a few days. Some of the compositional changes that occur in human milk are shown in Fig. 8.2. Finally, in some species a change in composition, especially in fat content, may occur during the course of suckling or milking.

It should now be clear how difficult it is to interpret data on milk composition. The dairy chemist works with milk samples from a bulked pool collected over a 24-hour period, but with most species this is not feasible; however, collecting a few millilitres of milk from a mammary gland is most unlikely to give a representative sample.

Attempts over the last century to formulate unifying concepts relating the composition of the milk of a species to the growth of its young have not been very satisfactory. Nutritional adequacy depends not only on quality, but also on quantity, and the range of species studied has generally been too limited to warrant generalization. However, there does appear to be some merit in the hypothesis proposed by Ben Shaul, namely that

Fig. 8.2. Changes in the concentrations of (*a*) lactose and protein, and (*b*) calcium and magnesium, in the mammary secretions of eleven women during the peri-partum period. The zero on the horizontal axis indicates time of delivery. (From J. K. Kulski and P. E. Hartmann. *Aust. J. Exp. Biol. Med. Sci.* **59**, 101–14 (1981).)

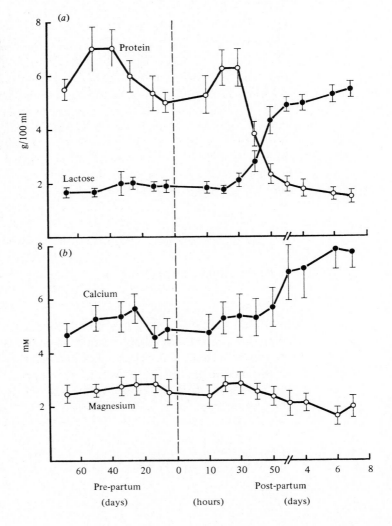

species that nurse frequently tend to produce milk with lower concentrations of nutrients than those that nurse their young infrequently. Certain arctic, aquatic and desert mammals have milks in which fat is the main constituent, providing 75 per cent or more of the energy value of the milk. In the arctic and aquatic animals the high fat content helps the young to offset heat loss, while in the desert species it helps to conserve maternal water.

In addition to their nutritional role, colostrum and milk have an important function as protective agents against infections. In ungulates and marsupials they are the sole routes of transfer of maternal antibodies to the young. Even in those species that are born with maternal antibodies in their blood received by way of the placenta (e.g. man, monkey, rabbit), antibodies in the milk and colostrum, although not absorbed, may provide an important protection against infections within the gut lumen.

Immunoglobulin A (IgA) is produced locally in the mammary gland by plasma cells. In the process of its secretion, dimers (double molecules) are formed, and a glycoprotein chain, termed a 'secretory component', becomes attached to the dimer to form secretory IgA (SIgA) which is more resistant to changes in pH and to proteolytic enzymes. There is now evidence that many of these plasma cells in the mammary gland are derived from sensitized B-lymphocytes that migrated from the Peyer's patches in the gut and are capable of secreting specific IgAs against microorganisms previously ingested by the mother. This entero-mammary transfer of lymphocytes and the release of their antibodies into the milk thus protects the suckling against pathogens in its immediate environment to which the mother has already become immune. In mice there is also evidence that oestrogen, progesterone and prolactin regulate the migration of lymphoid cells to the mammary gland. In addition to immunoglobulins, colostrum and milk can be rich sources of several substances that exhibit antimicrobial activity, e.g. lysozymes, lactoferrin and the lactoperoxidase system.

The nutritional and immunological roles of colostrum and milk are well recognized. We must now ask whether in some species these mammary secretions may not also exert hormonal effects in the young. A variety of hormones are present in both colostrum and milk (e.g. progesterone, oestrogens, prolactin). It may be that the mammary gland is just one of the excretory routes for the hormones, and that they are of no physiological significance to the infant. Nevertheless it appears that prolactin and a gonadotrophin-releasing hormone present in the milk of the rat enter the circulation of the neonate in biologically active forms. Further studies in this area are certainly required. Whatever the physiological significance of these hormones, assays for progesterone and oestrogen in milk are now exploited commercially as a reliable means of detecting early pregnancy in cattle (see Chapter 7).

The mammary gland

Gross anatomy

The number, shape and size of the mammary glands vary greatly in different species. They are present in both sexes although generally poorly developed in the male, and are normally functional only in the female. In most species the mammary glands are paired, varying from two, as in man, guinea pig and goat, to 14–18 as in the sow; in some marsupials the number is odd because two glands fuse at an early stage of fetal development. Some species have pairs of glands closely apposed in a structure termed an udder, like the two pairs in the cow, and one pair in the goat and sheep (Fig. 8.3). The position of the mammary glands varies – thoracic in man, elephant, monkey and the bat; extending along the whole length of the ventral thorax and abdomen in the sow, rat and rabbit; inguinal in the ruminants; abdominal in the whale, and almost dorsal in the coypu. As to shape and form there is great variety – in the rat the six glands on either side form relatively flat sheets of tissue enveloping the body wall; in the rabbit the glands are flat but circular in outline; they may be prominent as in man,

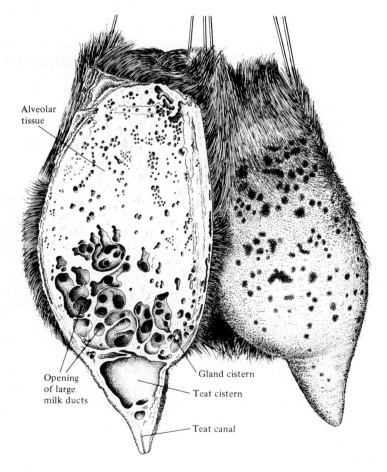

Fig. 8.3. Udder of a goat in which part of the left mammary gland has been cut away to show the dense alveolar tissues, the gland cistern with the large ducts opening into it, the teat cistern and the teat canal.

Alveolar tissue

Opening of large milk ducts

Gland cistern

Teat cistern

Teat canal

or dependent as in ruminants. In all female mammals, with the exception of the monotremes, a nipple or teat is present on each mammary gland; in the males of some species, such as the rat and mouse, nipples are absent.

Whatever the external shape of the mature mammary gland may be, its basic internal structure is the same in all species. There are two distinct types of tissue within the mammary gland: first the true glandular tissue or parenchyma, and secondly the supporting tissue or stroma. The parenchyma in the functional gland consists of a single layer of epithelial cells which are the milk-secreting alveolar cells (Fig. 8.4). The alveoli occur in clusters; each alveolus opens into a small duct and these small ducts join up to form larger ducts which eventually open to the exterior at the tip of the nipple or teat; in monotremes, as noted above, they open on special areas on the surface of the skin of the abdomen. While the basic structure of the mammary gland is similar, there is much species variation in the precise pattern of the duct system. In the rat and mouse the ducts eventually join to form one common duct or galactophore which leads directly through the nipple (Fig. 8.5a); in the rabbit the mammary ducts unite until there are some six to eight main ducts or galactophores, each draining a sector of the mammary gland, and these pass separately through the nipple (Fig. 8.5b). In some species certain parts of the ducts are dilated to form sinuses which act as storage spaces. In man, for example, some twelve to twenty galactophores pass through the nipple and each in the

Fig. 8.4. Diagram of a cluster of alveoli in the mammary gland of a goat.

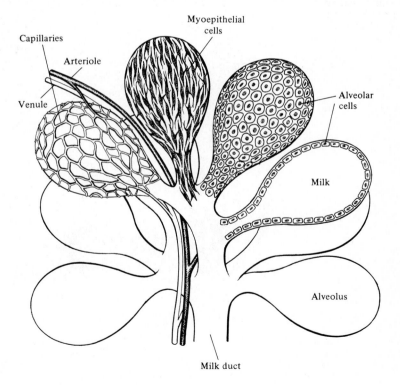

region of the base of the nipple, expands to form a sinus (Fig. 8.5*c*). In ruminants the mammary storage space is quite gross; the larger ducts terminate in a common gland cistern, a large cavity within the gland which leads directly into a smaller cistern within the teat; this in its turn leads to the teat or streak canal which opens at the tip of the teat (Fig. 8.5*d*).

The varied positions of the mammary glands in different species mean that their blood supply and nervous connections must also differ. Since the mammary gland is of cutaneous origin it shares the blood and nerve supply of the contiguous skin.

Microscopic structure

In the fully developed functional mammary gland the alveolar walls are formed by a single layer of epithelial cells whose shape varies with the amount of secretion being stored in the lumen of the alveolus; when the lumen is empty the cells are tall, but when the alveolus is full of secretion the cells are low and stretched (Fig. 8.6*a* and *b*). Overlying the base of the epithelial cells is a network of star-shaped myoepithelial cells which, because of the manner in which they envelop the alveolus, were at one time

Fig. 8.5. Diagram showing four different arrangements of the mammary duct system. (*a*) Rat: ducts all unite to form one main duct or galactophore (G) which opens at the tip of a nipple (N). (*b*) Rabbit: ducts unite to form several main ducts. (*c*) Woman: each main duct is dilated near the base of the teat to form a sinus (S); the nipple is surrounded by the dark-coloured areola (A). (*d*) Ruminant: the main ducts (MD) open into a single large gland cistern (GC) which in turn opens into the teat cistern (TC). TCL, teat canal.

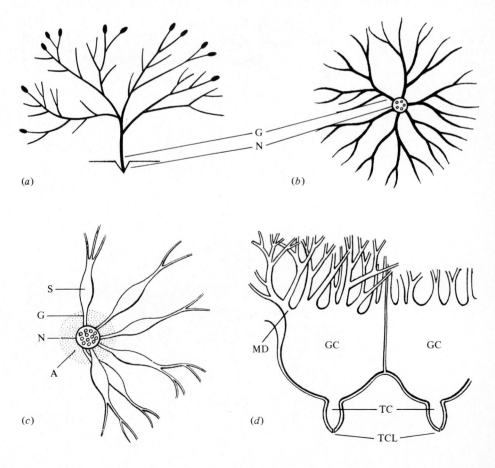

termed basket cells (Fig. 8.7). On top of the myoepithelium is a network of capillaries which supplies the alveolar cells with the necessary precursor substances for the synthesis of milk.

When examined under the electron microscope (Fig. 8.8), the alveolar cells in the functional gland are seen to have on their free (i.e. luminal) surface numerous fine projections, the microvilli. Each cell is firmly joined to its neighbour by junctional complexes just below the luminal surface; the lateral surfaces of alveolar cells are almost straight. The bases of the alveolar cells abut onto the myoepithelial cells or onto the basement membrane and are indented into a system of clefts which, by increasing

Fig. 8.6. Cells in the wall of an alveolus, (*a*) just after milking, (*b*) just before milking. As the alveolus fills up with milk and its walls are stretched, the shape of the cells is much altered. The empty capillaries at the base of the alveolar cells show up clearly because the tissue was fixed by intravascular perfusion. ((*a*) From S. J. Folley. In *Marshall's Physiology of Reproduction*, 3rd ed. vol. 2 Chap 20. Ed. A. S. Parkes. Longmans; London (1952); (*b*) from K. C. Richardson. *Proc. R. Soc. B*, **136**, 30 (1949).)

(*a*)

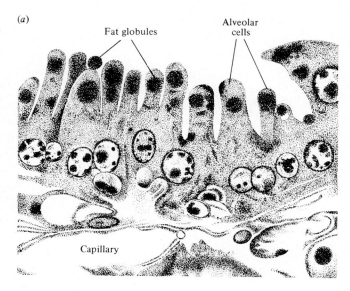

Fat globules

Alveolar cells

Capillary

(*b*)

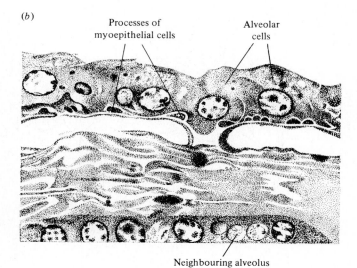

Processes of myoepithelial cells

Alveolar cells

Neighbouring alveolus

the surface area of the cells, probably facilitates absorption of milk precursors. The cell nucleus is large and rounded. A characteristic feature of the cytoplasm is the abundant endoplasmic reticulum consisting of cytoplasmic membranes arranged as arrays of flattened sacs orientated parallel to one another with their outer surfaces covered with numerous RNA–protein particles (ribosomes) and mostly situated in the basal two-thirds of the cell. The mitochondria are large with conspicuous partitions, or cristae. Another striking ultrastructural feature of the functional alveolar cell is the large Golgi apparatus, located in the more apical part of the cell, also adjacent to the nucleus, consisting of stacks of flattened sacs and vacuoles which, by their numbers, may give sections of that part of the cell a 'moth-eaten' appearance. The myoepithelial cells have a spindle-shaped nucleus, their cytoplasm is rarefied and contains fine muscle-like myofilaments which run parallel to the long axis of the cell's processes and are attached to the cell membrane. As noted later, these myoepithelial cells have a contractile function and are concerned in the expulsion of milk from the lumen of the alveolus into the duct system.

Growth of the mammary gland in the fetus

The evolution of the mammary gland has been much discussed and its origins variously associated with the primordia of sweat glands, sebaceous glands, hair follicles or, in birds, with the brood patches; however I do not propose to extend such speculations. The mammary glands are derived from the ectoderm in the embryo: two linear thickenings appear as a ridge, the milk line or crest, on either side of the mid-line. This ridge becomes interrupted into a series of nodules of ectodermal cells, the number and position of these depending on the species. These nodules, the early

Fig. 8.7. Myoepithelial cells on the outer surface of alveoli. (From a photograph by courtesy of Mr K. C. Richardson.)

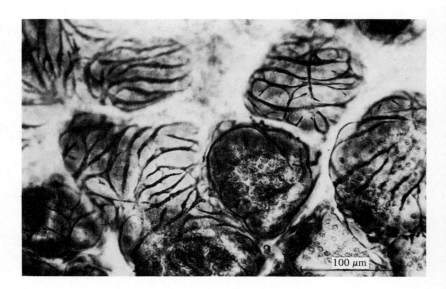

100 μm

Fig. 8.8. Diagram of the ultrastructure of three alveolar cells and a myoepithelial cell.

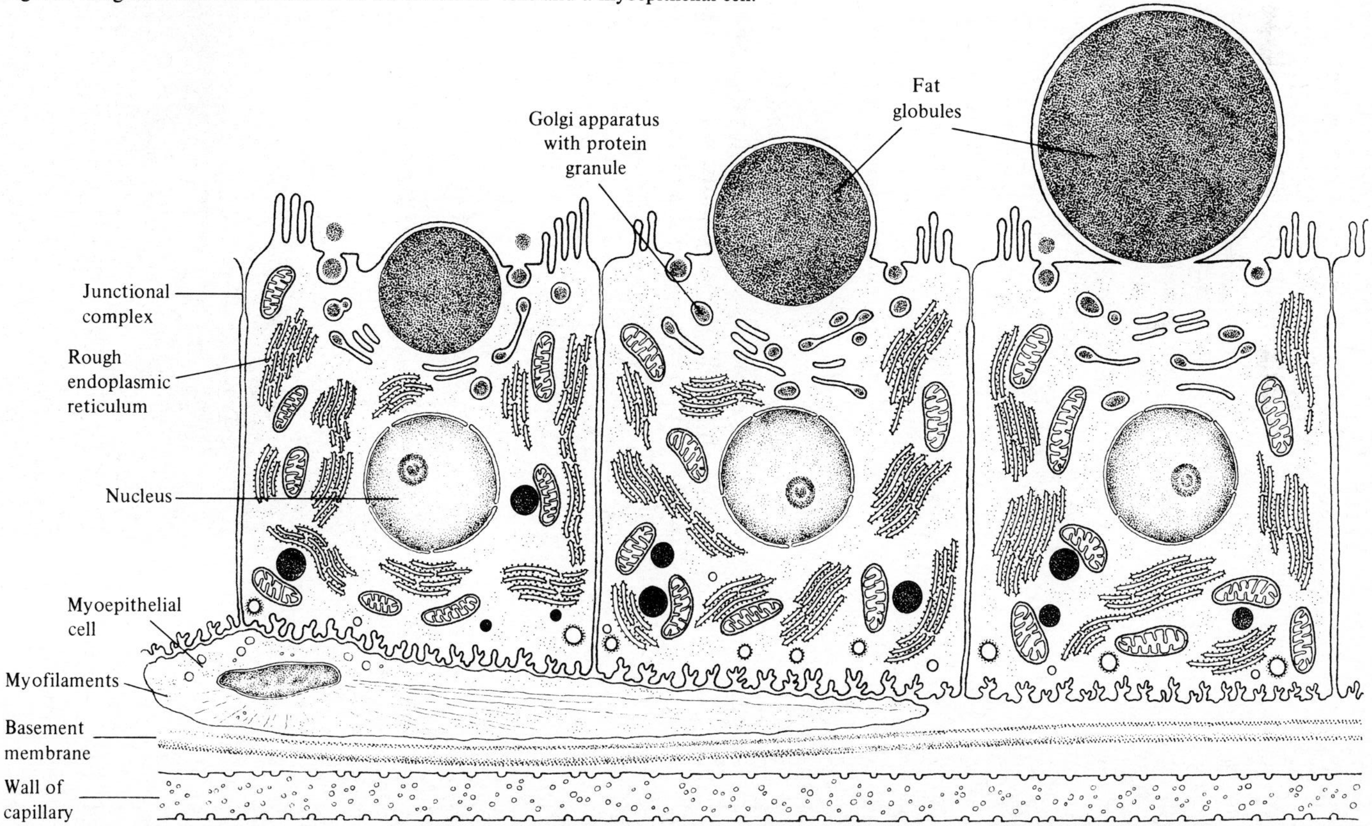

Fat
globules

Golgi apparatus
with protein
granule

Junctional
complex

Rough
endoplasmic
reticulum

Nucleus

Myoepithelial
cell

Myofilaments

Basement
membrane

Wall of
capillary

rudiments of the mammary glands, sink into the dermis to become mammary buds. At first lens-shaped, the bud becomes more spherical and later conical. A pause then ensues in the bud's growth after which it elongates into a cord-like structure, the primary mammary cord, whose base remains attached to the epidermis while its distal end penetrates into the dermis (see Fig. 8.9*a*). The distal end then branches into two, or may produce a number of secondary buds depending on the species. These buds elongate into cores which become hollow, forming mammary ducts whose walls are composed of two layers of cuboidal cells. In this fashion the mammary duct system is laid down in the developing fetus. In some species sex differences occur in the pattern of fetal mammary growth; for example, in the male rat and mouse in the second half of fetal life the primary cord loses its attachment to the epidermis (Fig. 8.9*c*), with the result that in the postnatal male the mammary ducts have no communication to the exterior and no nipples are formed.

Fig. 8.9. Development of the mammary gland in the fetal mouse and the effect of gonadectomy. (From A. Raynaud. In *Milk: the Mammary Gland and its Secretion*, vol. 1, Chap. 1. Ed. S. K. Kon and A. T. Cowie. Academic Press; New York and London (1961).)

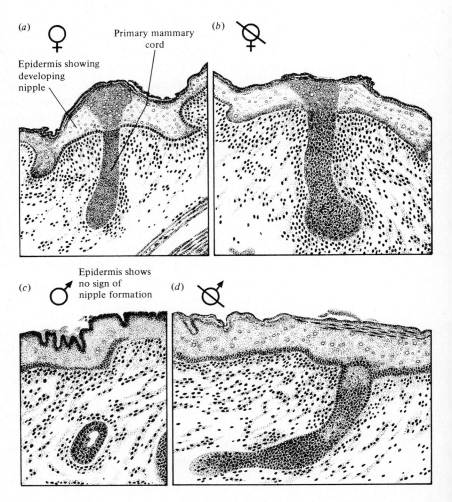

Postnatal growth of the mammary gland

At birth the mammary apparatus is represented by a rudimentary duct system leading to a small nipple. Until just before the onset of reproductive cycles in the female, there is little duct growth, for this is a quiescent phase in mammary development. As the onset of regular ovarian cycles approaches, a phase of active mammary growth begins, the nature of which varies with the species and tends to be related to the type of sex cycle. In species with short cycles, e.g. of 4 to 5 days as in the rat and mouse, the ducts grow and branch until they are fully extended. Alveoli are not usually formed but quite extensive alveolar formation does occur in the female hamster (4-day cycle). In species such as the primates, with a more prolonged luteal phase of the cycle, the branching of the duct system continues until the future lobules are indicated by collections of fine ductules and by the formation of some alveoli surrounded by a delicate stromal tissue. Finally, in those species in which the luteal phase of the cycle is prolonged into a pseudopregnancy, the duct growth and lobulo-alveolar formation proceed to a degree that is observed only in late pregnancy in the former two groups.

In the human female during puberty there is an extensive and character-istic growth of the stromal tissues of the mammary gland resulting in the formation of the prominent breast. We should note that the term 'mammary development' in animals means growth of the glandular tissues, i.e. ducts and alveoli; while the term may be used in this sense with the human mammary gland, it is generally used in the sense of growth of the stroma. This ambiguity is unfortunate since the overall size of the breast in the non-pregnant woman does not accurately reflect the amount of true glandular tissue within, or its lactational potential.

In monotremes during the breeding season the mammary ducts extend and branch in a similar fashion in 'pregnant' and 'non-pregnant' females; alveoli develop in response to incubation of eggs. In marsupials the duration of pregnancy is frequently shorter than one oestrous cycle and again the growth of the mammary ducts is similar in pregnant and non-pregnant females. If new-born young are fostered on the teats of a virgin oestrous female, mammary growth and lactation ensue just as in the pregnant female after parturition.

In most eutherian mammals full mammary growth is not achieved until the end of pregnancy or even early lactation. Usually not until mid-pregnancy (Fig. 8.10) is the full impetus of mammary growth observed; then there is a rapid formation of lobules of alveoli which take over much of the space formerly occupied by stromal tissue, so that by the last third of pregnancy the stroma is represented by narrow bands of connective tissue further dividing the gland into lobes. In some species during the last third of gestation the alveolar cells begin to secrete and the lumen of the alveolus becomes distended with a secretion containing fat globules. This is particularly striking in ruminants (Fig. 8.10); in other species, such as rats, secretory activity is not observed until just before parturition. In all species there is a further burst of secretory activity at the time of parturition or soon afterwards.

Control of fetal mammary growth

Information about factors that regulate mammary growth in the fetus is as yet limited to a few species. Early mammary differentiation is apparently

Fig. 8.10. Sections of the mammary gland of the goat at three different times during pregnancy (gestation is about 150 days). 35th day of pregnancy – note the small collections of ducts scattered throughout the stroma; 92nd day of pregnancy – the lobules of alveoli are now forming in groups known as lobes; secretion is present in some of the alveolar lumina and there is still quite a lot of stromal tissue; 120th day of pregnancy – the lobules of alveoli are almost fully developed; the alveoli are full of secretion and the stromal tissue is reduced to thin bands separating lobules and thicker strands between lobes. (35th day and 120th day from A. T. Cowie. In *Lactation*. Proceedings of an International Symposium forming the Seventeenth Easter School in Agricultural Science, University of Nottingham. Ed. I. R. Falconer. Butterworths; London (1970). 92nd Day – unpublished data of A.T.C.)

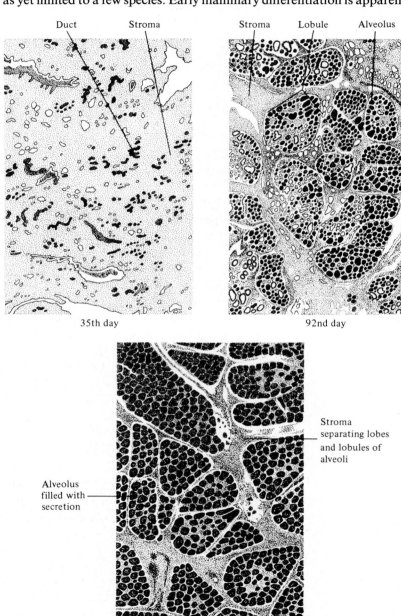

Duct   Stroma   Stroma   Lobule   Alveolus

35th day   92nd day

Stroma separating lobes and lobules of alveoli

Alveolus filled with secretion

120th day

not controlled by hormones since the mammary buds and primary cords will develop *in vitro* in explants of the ventral body wall (from mouse and rabbit embryos) placed in culture before the appearance of the mammary rudiments. Mesenchyme plays an important role in regulating the development of the mammary epithelium; this has been demonstrated by the ingenious studies of A. Propper and of K. Kratochwil. They separated the mammary epithelium and its mesenchyme by enzymic digestion; then, they investigated the mesenchyme's regulatory role by recombining mammary epithelia of different ages with mesenchyme from mammary and other regions in culture. Mammary epithelium isolated from the 12-day-old rabbit embryo fails to develop into a primary mammary bud unless it is combined with mammary mesenchyme; indeed at this stage the mammary mesenchyme will even induce bud formation in epithelium from a non-mammary region. By day 13 mammary epithelium will also respond to non-mammary mesenchyme. So potent is this inductive effect of mesenchyme that mesenchyme from a 13-day fetal rabbit, cultured in combination with epidermis from the flank of a 6-day chick embryo, will even induce the formation of structures resembling mammary buds! By day 8, however, the chick epidermis is committed to feather formation and no longer gives such a response. Since mammalian mesenchyme would thus seem to have the ability to induce mammary-like structure in avian epidermis, one may ask whether avian mesenchyme can induce feather growth in mammalian epidermis. Incredible as it may seem, if chick mesoderm is placed in contact with 14-day rabbit epithelium in which the mammary buds have already formed, the buds de-differentiate and slowly disappear, to be replaced by structures resembling rudimentary feather follicles!

In the fetal mouse the mammary epithelium is also under mesenchymal regulation, and the mesenchyme has an organ-specific effect; replacing mammary mesenchyme with salivary gland mesenchyme will induce the mammary epithelium of both fetal and postnatal mice to adopt the growth pattern of a salivary gland.

While early mammary differentiation appears to proceed in the absence of hormones, some of the later fetal stages are regulated by hormones. In the mouse and rat, as noted above, a sex difference occurs in the mammary development in the second half of fetal life (days 15–16) – in the male the primary cord disappears so that the mammary bud loses its attachment to the epidermis, and nipple formation is inhibited. This response is caused by androgens from the fetal testis; destruction of the testes on day 13 allows the male gland to develop as in the female. Destruction of the ovaries of the female fetus, however, does not influence the female pattern of growth, but injection of androgens into the female fetus will induce the male type of mammary changes. Thus in the mouse and rat the male pattern of mammary growth is a specialized type induced by androgens which modify the female (neutral) pattern of growth. Ingenious recombinant studies with mammary epithelium and mesenchyme from a strain of mouse that is androgen-insensitive (showing testicular feminization) indicate that these mammary changes in the male fetus are effected by a direct action of testosterone on the mesenchyme. Only mesenchyme from the mammary region responds to testosterone in this way and it must be from a species (e.g. mouse or rat) in which the mammary glands are sexually differentiated. In species like the rabbit in which the mammary glands are not sexually differentiated at birth, this is apparently because the mammary mesenchyme is not testosterone sensitive.

Evidence of hormone-mediated alterations to fetal mammary growth is lacking in other species, although it is certainly true that by the end of gestation the mammary gland of several species has acquired the capacity to respond to hormones, as revealed by the occurrence of precocious secretion – witch's milk – in the newborn. That fetal mammary growth may at times be affected by hormones is of clinical importance – some of the so-called spontaneous malformations of the mammary gland of the newborn may be associated with the effects of hormones administered to the mother during pregnancy. There is, moreover, evidence in rodents that perinatal hormone treatments may irrevocably alter the concentrations of the mammotrophic hormones or the mammary responsiveness to these hormones, or both, resulting eventually in abnormal mammary growth.

Control of postnatal mammary growth

Little is known about the control of postnatal mammary growth in either monotremes or marsupials, but it has been widely studied in some eutherian mammals. In these, mammary growth is largely controlled by hormones arising from the anterior pituitary gland, from the ovaries, and from the adrenal cortex; during pregnancy, the placenta is an additional source of both steroid and polypeptide hormones, but the relative importance of certain of these hormones appears to vary with the species. Of all the body organs none is more completely regulated by hormones in its growth and function than is the mammary gland.

In the rat and mouse, oestrogen from the ovary triggers off the sudden spurt of mammary duct growth just before the onset of oestrous cycles; ovariectomy abolishes the onset of this increased growth, which can then be restored by injection of suitable doses of oestrogen. Growth of the lobulo-alveolar system usually requires both oestrogen and progesterone – produced by the ovaries, by the adrenal cortex or by the placenta – although in some species oestrogen alone, if given in suitable doses, will induce lobulo-alveolar growth. This, however, is only part of the story, for these steroid hormones have little or no effect on mammary growth if the anterior pituitary is removed. Research shows that the steroids affect mammary growth in two ways, first by releasing prolactin (and probably other hormones) from the anterior pituitary, and secondly by acting directly on the mammary parenchyma; the latter action is only possible however if anterior pituitary hormones are also present. By removing the ovaries, adrenals and pituitary from infantile rats, the minimal hormonal requirements for mammary growth in these 'endocrinectomized' animals can be determined by studying the effects of various combinations of hormones. For duct growth, oestrogen, growth hormone and adrenal steroids are necessary; lobulo-alveolar growth will ensue if prolactin and progesterone are also added to this triad (Fig. 8.11). To what extent these observations apply to other species is still uncertain, but in the hypophysectomized goat a preliminary study has suggested that all five hormones are required for lobulo-alveolar growth. Prolactin-like hormones are present in the placenta of primates, ruminants and rodents, as well as some other species, and these placental hormones appear to have important roles in mammary growth during pregnancy.

Lactation

Lactation involves two distinct phenomena or phases, and recognition of these is all-important for understanding the physiology of lactation; their components are set out in Table 8.1.

Table 8.1. *The phases and components of lactation*

Lactation	
1. Milk secretion	2. Milk removal
Synthesis of milk constituents within alveolar cell	Passive removal of milk from cisterns and large ducts
Intracellular transport of constituents	Reflex ejection of milk from alveoli ('let-down', 'draught')
Discharge of constituents from cell into alveolar lumen	

Fig. 8.11. The hormones concerned in the growth of the mammary gland and in the initiation of milk secretion in the hypophysectomized-ovariectomized-adrenalectomized rat. (From W. R. Lyons. *Proc. R. Soc. B*, **149**, 303 (1958).)

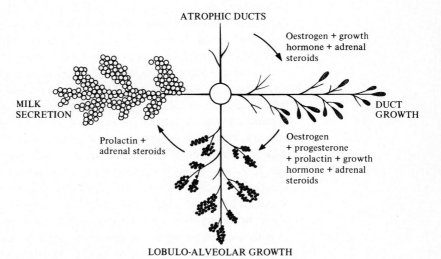

ATROPHIC DUCTS

Oestrogen + growth hormone + adrenal steroids

MILK SECRETION

DUCT GROWTH

Prolactin + adrenal steroids

Oestrogen + progesterone + prolactin + growth hormone + adrenal steroids

LOBULO-ALVEOLAR GROWTH

Milk secretion
The alveolar cells synthesize within their cytoplasm the constituents of milk from substances taken up from the blood, and then pass them out into the lumen of the alveolus; however it was not until the advent of the electron microscope with its high resolving power that we reached a reasonable understanding of the cellular processes involved in milk secretion.

Milk fat is mainly triglyceride, the fatty acids being synthesized in the gland or derived from the diet. There is still uncertainty about the precise location of lipid accumulation into droplets that occurs in the basal cytoplasm of the alveolar cell. On reaching the apex of the cell (see Fig. 8.8) the fat droplet pushes against the cell membrane which loses its microvilli and bulges into the alveolar lumen; the cell membrane fits tightly over the droplet and gradually constricts behind it so that a narrowing neck of cell membrane is formed between the enveloped droplet and the cell apex. The fusion of Golgi vesicles in the region of the fat droplet may aid the extrusion process. The narrowing neck of cell membrane finally becomes pinched off and the fat droplet, enclosed in cell membrane, falls free into the lumen. Sometimes portions of the alveolar cell cytoplasm get pinched off and enveloped with the fat droplet.

Milk proteins are synthesized from amino acids taken up by and synthesized in the mammary gland. After their synthesis on the endoplasmic reticulum the milk proteins pass to the Golgi apparatus where the caseins undergo phosphorylation and form micelles within the vesicles. The Golgi vesicles and their contents move towards the cell apex and fuse with the cell membrane; the fused membranes then form an opening through which the contents are discharged into the alveolar lumen.

Lactose is synthesized from glucose also within the Golgi vesicles, in association with the milk protein; indeed there is an interesting relationship between lactose and protein. The final step in the biosynthetic pathway of lactose is accomplished by an enzyme called lactose synthase, which is made up of two protein components, one of which is the milk protein α-lactalbumin. α-Lactalbumin is believed to be synthesized within the membranes of the rough endoplasmic reticulum and then passed into the Golgi apparatus where the second component of the enzyme – galactosyltransferase – is probably attached to the Golgi membranes. The conjunction of the two components of lactose synthase thus permits the completion of the synthesis of lactose which then passes into the lumen, probably with the protein granules. The synthesis of lactose is thus dependent on the prior synthesis of the milk protein α-lactalbumin.

Hormonal control of milk secretion
As we have seen, milk secretion begins in some species during the last third of gestation, whereas in others the onset of secretory activity is more closely associated with parturition. The secretion formed at this time is

called colostrum and differs in composition from normal milk. Soon after parturition there is a greatly increased secretory activity, and the colostral type of secretion changes to the formation of ever-increasing amounts of milk of normal composition.

How is the onset and the maintenance of secretion controlled? Various studies indicate that secretory nerves are not concerned. These include some striking experiments by the late Jim Linzell in goats, in which the mammary gland was transplanted to the neck and separated from its nerve

Fig. 8.12. Mammary gland transplanted to the neck of a goat, showing that an intact nerve supply is not necessary for normal lactation.

supply without subsequent loss of its ability to secrete milk (Fig. 8.12). It thus seems that the control mechanisms are hormonal rather than neural.

This hormonal regulation of milk secretion has been investigated by several approaches, including the classical one of removing an endocrine gland followed by replacement therapy. Removal or destruction of the anterior pituitary causes an immediate decline and ultimately a cessation

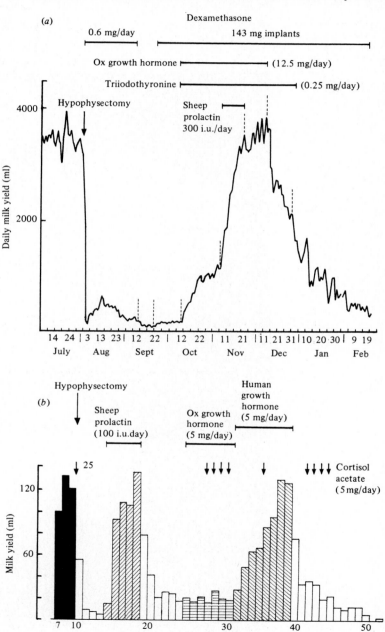

Fig. 8.13. Diagrams of daily milk yields of (*a*) goat and (*b*) rabbit, hypophysectomized during lactation and then treated with hormones to restore lactation. ((*a*) From A. T. Cowie. In *Lactogenesis: the Initiation of Milk Secretion at Parturition*. Ed. M. Reynolds and S. Folley. University of Pennsylvania Press; Philadelphia (1969). (*b*) From A. T. Cowie, P. E. Hartmann and A. Turvey. *J. Endocr.* **43**, 651 (1969).)

Fig. 8.14. Average milk yields of three control goats (closed circles) and three goats treated with increasing doses of bromocriptine (CB 154) (open circles). S1–S4, blood samples collected. In the bromocriptine-treated group, the blood levels of prolactin fell from 465 ± 57 ng/ml at S1 to 6 ± 0.4 ng/ml at S4, while that of growth hormone rose from 27 ± 9 ng/ml at S1 to 64 ± 11 ng/ml at S4. (By courtesy of Dr I. C. Hart.)

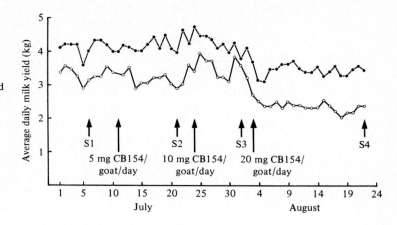

Fig. 8.15. (*a*) Average daily milk yield, and (*b*) average percentage change in body weight, in seven low-yielding cows (broken line) and eight high-yielding cows (solid line). Arrows indicate blood sampling. (By courtesy of Dr I. C. Hart.)

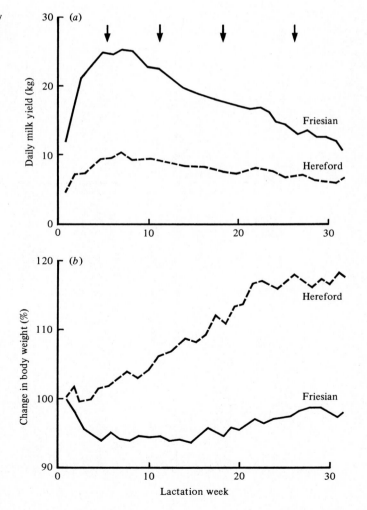

of milk secretion in all species. Replacement therapies have revealed that the hormones involved vary with species. In the lactating rat both prolactin and adrenocorticotrophin are necessary for the maintenance of milk secretion after hypophysectomy; in the rabbit, prolactin by itself is sufficient (Fig. 8.13*b*). In the goat and sheep, prolactin, growth hormone, adrenocorticotrophin and thyrotrophin are all required to restore a high milk production but once this has been achieved, high yields may continue in the absence of prolactin (Fig. 8.13*a*). Complementary information has come from the responses of lactating animals treated with dopamine agonists (e.g bromocriptine), which specifically inhibit the release of prolactin from the anterior pituitary. In lactating non-ruminants such drugs rapidly inhibit milk secretion; in lactating ruminants they may have little or no effect on milk yield although levels of prolactin in the blood are much depressed (Fig. 8.14). If however, such drugs are given to

Fig. 8.16. Average plasma concentrations of (*a*) growth hormone, (*b*) insulin and (*c*) prolactin found throughout selected 24-hour periods in seven low-yielding cows (shaded columns) and eight high-yielding cows (open columns). (By courtesy of Dr I. C. Hart.)

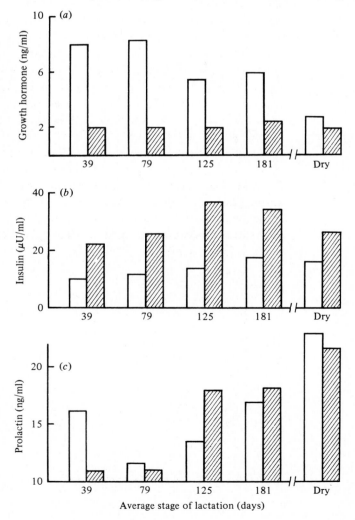

ruminants at about the time of parturition, the onset of copious lactation is largely inhibited.

In the last decade new sensitive assay procedures (such as radioimmunoassays and radioreceptor assays) have permitted detailed studies of the hormone concentrations in the blood during lactation as well as of the behaviour of the hormone receptors in the mammary gland. Cows can differ in the manner in which they partition their dietary energy between milk production and body growth (see Fig. 8.15). Radioimmunoassays have revealed that high milk-producing cows have higher concentrations of growth hormone but less insulin in their blood than do the beef type that readily put on flesh; their levels of prolactin however do not differ (see Fig. 8.16). In ruminants the concentration of prolactin in the blood varies greatly with the seasons of the year, being high in summer and low in winter. These seasonal changes are induced by changes in day-length and temperature (see Fig. 8.17); there is no clear evidence that they affect lactation. (The high levels or prolactin in the dry period in Fig. 8.16 are summer values.) Seasonal changes in the levels of circulating prolactin also occur in male ruminants; in male sheep and goats mid-summer concentrations may be some thirty times higher than mid-winter values.

In summary, although the maintenance of lactation usually requires the presence of several anterior pituitary hormones, prolactin has a critical role in all non-ruminant species so far studied; in ruminants the major role is played by growth hormone.

The precise hormonal mechanisms involved in the initiation of copious milk secretion around parturition have still to be completely elucidated. The differences in times of onset of secretory activity in the mammary gland during gestation in different species suggest that variations may well exist in the mechanisms. In the pregnant rat the lactogenic activity of pituitary prolactin and placental lactogen is blocked by a direct action of progesterone on the alveolar epithelium. Some 30 hours before parturition there is a

Fig. 8.17. Seasonal variations of prolactin concentrations under normal variations of daylight in Ile-de-France ewes. (Modified from J. Thimonier, J. P. Ravault and R. Ortavant. *Ann. Biol. Anim. Biochem. Biophys.* **18**, 229–35 (1978).)

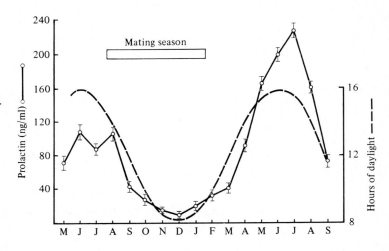

switch in steroid production by the ovaries and a resulting drop in the level of progesterone in the blood which permits the lactogenic effects of prolactin, placental lactogen and adrenal steroids to come into play. In the rat and mouse there is some evidence that progesterone exerts part of its inhibitory effect by blocking the synthesis of α-lactalbumin, thereby inhibiting the synthesis of lactose.

In other species the mechanism has been less clearly defined, and while progesterone withdrawal would seem to be an important factor in most eutherian mammals, falling progesterone levels can only trigger copious milk secretion in the presence of adequate levels of lactogenic hormone, in particular, prolactin.

Milk removal

The second phase of lactation is concerned with the transport of the stored milk from the alveolar lumina to the nipple or teat where it can be readily removed by the infant (or milker). In species in which there are mammary cisterns or sinuses, the milk in them is immediately available to the infant, and its removal is said to be passive; removal of the greater portion of the milk from the alveoli, however, requires the triggering of a neurohormonal reflex – the milk ejection reflex – which is discussed at length in Chapter 2.

A minute or so after the infant is put to the breast, or the milker starts to milk the cow, the mammary glands swell and become tense with milk under pressure (see Fig. 8.18), which may spurt or leak from the nipples

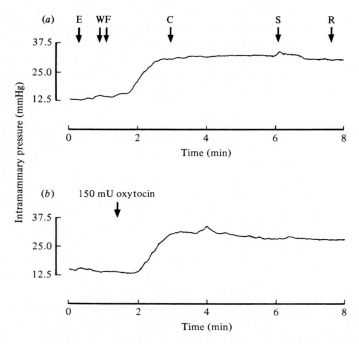

Fig. 8.18. Intramammary pressure recorded from an unmilked rear quarter in a cow (a) during machine milking and (b) during a rapid intravenous injection of 150 mU oxytocin. E, entry of milker; W, washing of udder; F, fore-milking; C, application of teat cups; S, stripping; R, removal of teat cups. (From J. D. Cleverley and S. J. Folley. *J. Endocr.* **46**, 347 (1970).)

or teats. This phenomenon, colloquially called the 'draught' in women and 'let-down of milk' in cows, has long been recognized. It can become a conditioned reflex and occur in anticipation of the suckling or milking stimulus. On the other hand, discomfort or any form of stress can inhibit it so that much less milk is obtained by the child or by the milker. For many years the sudden milk flow was attributed to an increase in the rate of milk secretion. However, it was shown that the cow's udder contains all the milk obtainable on milking *before* the occurrence of the 'let-down', and there is now ample evidence that the rapid increase in intramammary pressure is due solely to the sudden expulsion of the milk from the alveoli and fine ducts into the large ducts, sinuses or cisterns, and not to increased secretion. This movement of milk is brought about by contraction of the myoepithelial cells squeezing the alveoli and expelling their contents, the phenomenon now termed milk ejection. Extensive investigations have revealed that the act of suckling or milking triggers nerve impulses from receptors in the nipple or teat, which pass up the spinal cord to the hypothalamus where they cause the release of the hormone oxytocin from the posterior lobe of the pituitary gland into the blood stream. On reaching the mammary gland, oxytocin causes the myoepithelial cells to contract, thus increasing intramammary pressure (Fig. 8.18). The milk ejection response is a reflex; like ordinary reflexes it can be conditioned, but unlike ordinary reflexes its efferent arc is not nervous but hormonal.

Throughout the mammals there is a great range in the frequency of suckling, from once a week in some marine mammals to 50–80 times a day in the rat. Studied in detail by Dennis Lincoln and his colleagues, the pattern in the rat is indeed unusual (see Chapter 2). After attachment of the young to the nipples there may be a delay of some 10 minutes before milk ejection occurs, but thereafter milk ejection occurs spontaneously every 3 to 10 minutes as long as the young remain attached to the nipples. The importance of the proper functioning of this reflex seems to vary with species; in the rat and rabbit the young will get practically no milk even from full mammary glands unless milk ejection occurs; at the other extreme one can get milk out of the udder of a goat even in the total absence of the milk ejection reflex. This divergence may well reflect differences in the architecture of the mammary gland; there are no large storage sinuses or cisterns in the mammary glands of the rat or rabbit whereas there is a large gland cistern in the goat in which the secretion accumulates. In some species the myoepithelial cells may also respond to direct mechanical stimulation – witness the way the young lamb or calf 'butts' the udder. This may also aid milk ejection.

The concentration of oxytocin in the blood of cows and goats during milking and suckling has been measured by sensitive bioassay and radioimmunoassay procedures (see Fig. 8.19); the former give somewhat higher values but the patterns of release obtained by both assay procedures are similar. Comparison of the concentrations of the hormone in different species must be made with caution because the site of blood sampling may greatly affect the results. In ruminants blood for assay is usually collected from the convenient large external jugular vein which contains oxytocin-rich venous blood draining the posterior pituitary. In women, monkeys and pigs the posterior pituitary drains mainly into the jugular vein which is much less accessible, and blood is much more likely to be obtained from a peripheral vein that is easy to get at. However this means that any oxytocin present will already have passed around the pulmonary and systemic circulations and will therefore have been excessively diluted and perhaps inactivated. Failure to recognize the significance of these anatomical differences in the venous drainage of the posterior pituitary and the site of blood collection can lead to erroneous conclusions about apparent species differences in levels of oxytocin.

In the modern dairy cow the release of oxytocin and milk ejection generally occur soon after the teat cups are applied; indeed they may occur before the application as a conditioned response to being fed before milking or to the sight or sound of the approaching milker (see Fig. 8.18). Primitive breeds of cattle are less accommodating, and milk ejection does not readily occur in them in the absence of the suckling calf. To aid in the milking of his domestic animals, early man adopted two main subterfuges which exploited certain aspects of the milk ejection reflex. The obvious trick of bringing the calf to the cow and even allowing it to suckle one teat while the others were milked was frequently used; sometimes a dummy calf or even a boy covered with a calf's skin was used to elicit the reflex. Another procedure was to blow air into the vagina; we now know that vaginal distension stimulates oxytocin release, the so-called Ferguson reflex. This genital stimulation procedure was apparently known to the ancient Egyptians, and Herodotus described how the Scythians used pipes made of bone to blow into the vagina of their mares before milking. Indeed the practice of blowing air into the vaginae of cows to aid milking has been

Fig. 8.19. Oxytocin in the blood of cows before, during and after milking. PM, preparation for milking; MA, application of teat cups; S, stripping; EM, end of machine milking. C, control level. Abscissae show time in minutes. (From Schams *et al. Acta Endocrinologica* **92**, 258–270 (1979).)

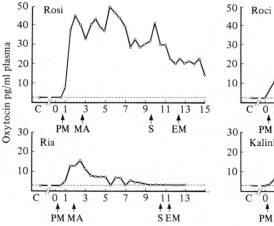
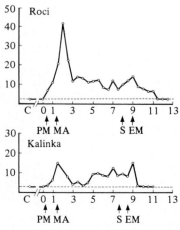

widely used in Europe, Asia and Africa (see Fig. 8.20) and is still used by some African tribes today. The widespread use in several continents, and over several millenia, of this technique to facilitate milking may at first sight seem remarkable and puzzling. Perhaps primitive man, observing on occasions the spurting of milk from the nipples of his lactating partner during sexual intercourse, turned the thought over in his mind that if vaginal stimulation induced 'the draught' in women it might also induce 'let-down' in animals. In this context, it is interesting that recent studies have been able to detect oxytocin in the blood of women after orgasm.

There existed a widespread belief that milk expulsion in marsupials and in marine mammals was aided by voluntary contractions of adjacent body muscles. In some specific instances this has now been shown to be false, and there would seem to be good reasons for believing that the neuro-endocrine reflex of milk ejection occurs in most, if not all mammals, including the monotremes, and that it is the principal mechanism for the transfer of milk from the alveoli to the large ducts and sinuses. Indeed, in the wallaby at least, the reflex has been refined to cope with the great differences in the amount of stimulation received between the small gland, being constantly suckled by the continuously attached young of less than 1 g body weight, and the large gland, being suckled intermittently by the young at foot weighing over 2500 g. The myoepithelia of the small gland are much more sensitive to oxytocin, thereby allowing the joey to suckle at will without 'turning on the taps' and flooding the pouch with milk from the large gland at each feed.

Fig. 8.20. A Hottentot blows into the vagina of a cow. (From *The Present State of the Cape of Good Hope*. P. Kolben. [Translated by Medley.] Innys; London (1731).)

Their Method to bring a Refractory Cow to yield her Milk.

Physiology of suckling

The new-born is guided to the nipple or teat by visual, olfactory and tactile cues (their relative importance depending on the species), and these may be supplemented by maternal coaxing. For the new-born rat, olfactory cues are all-important and are associated with traces of maternal saliva and amniotic fluid on the nipples. Maintenance of the already established act of suckling, however, depends on continued experience, and any interference on days 3 and 4 with the normal pattern of nipple search and attachment can seriously affect further suckling behaviour of rat pups.

It is widely believed that the young on the nipples or teats obtain milk only by suction, but they do not. Only young monotremes suck up milk but their mothers possess neither nipples nor teats! In eutherian mammals suction aids the process of obtaining milk but it is not an essential component. Cineradiographic studies in young babies and animals during suckling clearly reveal that milk is obtained by expressing it from the nipple or teat. The base of the teat is compressed between the tongue and hard palate, and the milk contained within is then stripped out of the teat by the tongue compressing the teat from its base towards its tip against the hard palate. The pressure on the base of the teat is then released to allow it to fill up with milk, which quickly happens as the milk is under pressure from the milk ejection reflex, and the whole action is repeated. The human

Fig. 8.21. A double-chambered teat cup. (a) Pulsation chamber is open to the atmosphere, i.e. a pressure of about 100 kilopascals (760 mmHg); within the liner and milk tube there is a relatively constant vacuum of 50 kilopascals (380 mmHg), so that the rubber liner collapses in onto itself below the tip of the teat, occluding the milk tube and hence the flow of milk. (b) Pulsation chamber connected to vacuum system, i.e. 50 kilopascals; the rubber liner is now fully dilated, the vacuum is restored to the tip of the teat, and milk flows into the milk tube. (By courtesy of Mr D. N. Akam.)

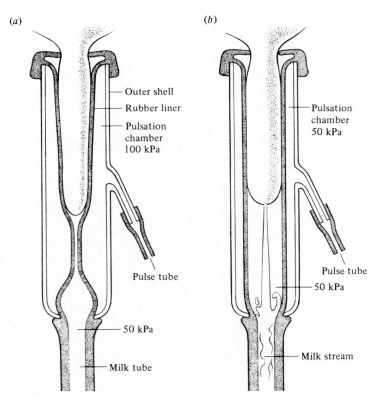

(a)

Outer shell
Rubber liner
Pulsation chamber 100 kPa

Pulse tube

50 kPa

Milk tube

(b)

Pulsation chamber 50 kPa

Pulse tube
50 kPa

Milk stream

infant forms a 'teat' by drawing the whole nipple and part of the areola into its mouth, but otherwise the procedure is the same. Young animals can very soon acquire the technique of obtaining milk by suction if fed from bottles with hard teats, but this is not their normal method.

It is of interest to note that the milking of cows by hand involves an action similar to that of the calf's tongue. The base of the teat is compressed between the thumb and first finger thereby trapping the milk in the teat cistern; a progressive compression of the teat by the second, third and fourth fingers then forces the milk out through the teat canal. Attempts to develop machines that would mimic the action of hand milking began a century ago, but these early machines were cumbersome and unsatisfactory. Modern machines obtain milk by suction; single-chambered suction cups were used originally, but the continuous vacuum applied to the whole teat caused much discomfort to the cow. Machines now have double-chambered teat-cups which have an outer metal shell fitted with a flexible rubber liner into which the teat fits; the lower end of the liner is continued as the milking tube which carries the milk away. The space formed between the rubber liner and the outer metal shell is the pulsation chamber, which is alternately connected to the vacuum system of the machine and to the atmosphere by a device termed the pulsator (see Fig. 8.21). When the pulsation chamber is connected to the atmosphere, the vacuum in the milking tube collapses the liner about the teat, compressing the lower part of the teat and stopping the flow of milk. As the vacuum is restored in the pulsation chamber, the liner is pulled away from the teat and milk flows again. The milk flow is thus intermittent. The rhythmic movement of the liner reduces the blood engorgement in the teat wall that would result from constant suction, and discomfort to the cow is thus avoided.

Suckling stimulus

It is now well established that in addition to the milk ejection reflex, the suckling or milking stimulus evokes a rapid reflex release of prolactin. In women this response tends to be greatest during the first two months of lactation; subsequently it may decline or become absent, although lactation continues. It seems that the magnitude of the prolactin release is determined by the frequency and the intensity of the suckling stimulus, which also plays a major role in suppressing ovulation in breast-feeding women (see Chapter 6).

The stimulus of suckling can thus release both prolactin from the anterior pituitary and oxytocin from the posterior pituitary, but they need not be released together. Are the nervous pathways common? In the spinal cord they probably are but there is little precise information. In the brainstem, however, much information has been obtained in several species by John Tindal and Geoffrey Knaggs, and, at the midbrain, the prolactin and oxytocin release pathways appear to be similar and to be associated with the spinothalamic system. Beyond this the prolactin-release pathway proceeds to the medial anterior hypothalamus, being joined by a descending pathway from the orbitofrontal cortex. It is believed that impulses following this pathway release prolactin by inhibiting transmission in the final neural link, so blocking the release of the prolactin-inhibiting factor, dopamine, into the hypophysial portal system (see Chapter 1). From the midbrain the oxytocin-release pathway proceeds to the medial forebrain bundle in the hypothalamus, from which collaterals make contact with the paired paraventricular and supraoptic nuclei whose neurones send axons to the posterior lobe of the pituitary. The complex role of these neurones in the transport and release of oxytocin is described in detail by Dennis Lincoln in Chapter 2.

The suckling or milking stimulus, being necessary both for milk secretion and for milk ejection, is thus of great importance in the general maintenance of lactation. Indeed in some species, such as the cat and rat, it appears to be essential; however, in the goat, at least under experimental conditions, it is apparently dispensable, for lactation can continue when all nervous connections between mammary gland and central nervous systems are destroyed, as in the mammary transplantation experiment referred to above. Other factors must therefore affect the release of anterior pituitary hormones; for example changes in blood composition, brought about by the uptake of milk precursor substances by the mammary gland, may be detected by special centres in the hypothalamus, thus causing a metabolic release of prolactin. Moreover, there is now evidence that areas of the brain much 'higher' than the hypothalamus participate in controlling the release of anterior pituitary hormones, so that the control of prolactin secretion may be quite complex.

Not only does the suckling stimulus maintain the supply of food to the offspring during its infancy but in some species the suckling stimulus can inhibit ovulation for prolonged periods of time; this biological contraceptive results in wide birth spacings, thereby allowing the population to remain relatively stable and to live within the nutritional resources of the habitat. Such a response to the suckling stimulus has been of particular importance in the evolution of the larger mammals, and especially of man himself. Interference with reproduction by stimulation of the nerve endings in the nipple seems to be mediated in part by raised levels of prolactin, and in part by inhibition of hypothalamic GnRH release. Roger Short has recently pointed out that lactation represents a key reproductive strategy in the Old World menstruating primates and in man. Recent demographic studies of one of the last remaining groups of *Homo sapiens* who still live as hunter–gatherers, the nomadic !Kung of the Kalahari Desert in South Africa, reveal that although the !Kung use no artificial forms of contraception and have no significant taboos on intercourse during lactation, their birth interval on average is just over four years. Babies live in close contact with their mothers and are breast-fed for short periods several times an hour throughout the day; they also sleep beside their mothers, and breast-feed while she sleeps. The pattern continues for the first three or four years of the child's life and is undoubtedly responsible for the long period of ovulatory inhibition. This, coupled with a relatively high infant mortality rate, results in a population doubling time of about 300 years. In societies who have abandoned frequent breast-feeding, the population may double in a generation! Indeed as Short remarks 'the changing history of breast-feeding is the history of the human population explosion'. In developing countries today breast-feeding still prevents more pregnancies than all artificial forms of contraception put together, and so it has a most important demographic impact. It is necessary to encourage frequent breast-feeding, ideally on demand, in order to maximize the contraceptive effect (see Chapter 6).

Clinical studies and uses of hormones in lactation
Hormone preparations have been used both in man and in animals to initiate, to increase or to suppress lactation.

Induction of mammary growth and lactation
In the late 1930s when synthetic oestrogens, such as diethylstilboestrol, first became available, the feasibility of using them to stimulate udder growth in ruminants was investigated, and these studies were extended during the Second World War with the object of bringing sterile cows and virgin heifers into lactation. The usual procedure was to implant tablets of oestrogen beneath the skin; a few weeks after implantation, 'milking' was started. Initially only a few drops of secretion were obtained, but gradually the volume increased and within a month the daily milk yield from a treated cow might be about 5 kg. After 6–10 weeks the unabsorbed portions of the oestrogen implant would be removed, when further increases in yield might occur, with yields of up to 13–14 kg of milk per day. However, treated animals often failed to give a satisfactory response, and because of this and other troublesome side effects, such as persistent oestrous behaviour, the procedure was never widely used. Subsequently, when synthetic progesterone became available, numerous studies were made on the induction of udder growth in goats with oestrogens and progesterone. This combination of hormones stimulated a type of lobulo-alveolar growth that was more normal in structure than that observed with oestrogen alone, when the alveoli had tended to be unduly large and even cystic.

In virgin goats treated with oestrogen plus progesterone to induce lactation, milk yields were about two-thirds of those expected had the animal been pregnant and come into lactation. These studies in goats also revealed the important contribution of the milking stimulus to the mammary growth process. Indeed, virgin goats can be brought into lactation by regular 'milking' without giving either oestrogen or progesterone. These experimental observations are, of course, in line with the numerous clinical reports of lactation occurring in non-pregnant women in response to repeated application of the suckling stimulus. As already noted, the ovarian hormones exert much of their mammary growth effects by releasing the appropriate hormones from the anterior pituitary, and the milking stimulus clearly enhances this release.

Interest in the induction of lactation in cattle revived in the last decade when it was observed that a short period of treatment with steroids – the daily injection of progesterone and a relatively high dose of oestradiol over a period of a week – induced mammary growth and lactation if milking was begun soon after the injections ended. The responses, however, have remained variable. Seldom can full growth of the mammary gland comparable to that occurring in late pregnancy be obtained with oestrogen–progesterone therapy, probably because of the absence of placental hormones in such studies.

Oestrogen therapy, including local application of a suitable cream, is sometimes used in the treatment of inadequate breast development in women or male transsexuals. This may produce a cosmetic effect in terms of the development of stromal tissue, but in women there may also be undesired side effects such as disturbance of the menstrual cycle. There was even a recent report of a husband who developed gynaecomastia because his wife had applied the oestrogen cream to herself either somewhat too lavishly, or had failed to rub it in thoroughly. This report is not without its ironical twist as the treatment was started because it was the husband himself who was dissatisfied with his wife's 'miserable little breasts'; moreover it illustrates that in some species the male mammary gland can respond quite readily to mammogenic hormones. It is not unknown for some male mammals, such as billy goats, to lactate spontaneously, without any apparent impairment of their libido or fertility.

Augmentation of an existing lactation
The milk yield of cows in declining lactation can be increased by suitable hormone therapy. Bovine prolactin is ineffective for this purpose. Bovine growth hormone, on the other hand, is effective, and the increases in milk yield are related to the dose of growth hormone administered over a fairly wide dose range (Fig. 8.22). However, until synthetic growth hormone becomes available this therapy cannot be used on a commercial scale.

Thyroid hormones have been used commercially to augment the milk yield of cattle. Being active by mouth, these can be incorporated into the food and give considerable increases in milk yield, but they must not be given for any length of time because of overstimulation of the general body metabolism; unfortunately when the treatment is stopped there is a dramatic decline in milk yield which reduces the overall gain to a mere 3 per cent. Thyroid hormone therapy is now therefore seldom used in cows.

Human prolactin, isolated in 1971, is not yet available in quantites sufficient to treat hypogalactia in women, but beneficial effects have been claimed in response to injections of thyrotrophin releasing hormone which stimulates the secretion of endogenous prolactin. The complex problems of failure of lactation in women are discussed later.

Inhibition of lactation

Oestrogens have been used to suppress unwanted lactation in women but their efficiency and safety have been much questioned. The suppression of lactation is now readily achieved by the use of dopamine agonists such as bromocriptine which inhibit the release of prolactin. In ruminants, however, these compounds only depress milk secretion when administered around parturition. In women, it is becoming increasingly apparent that the combined oral contraceptive pill has a significant inhibitory effect on lactation, so that its use is contraindicated in breast-feeding women. This is particularly important in developing countries, where the health of the infant may be critically dependent on an adequate supply of breast milk. Fortunately, gestogen-only contraceptives, like the 'minipill', or Depo-provera injections, do not have significant inhibitory effects on lactation.

Mammary cancer

Consideration of mammary pathology is beyond the scope of this chapter but since oestrogen appears to be involved in the aetiology of mammary cancer, brief reference to this condition may be warranted. Mammary cancer is extremely rare in species such as cows, sheep and pigs, but very

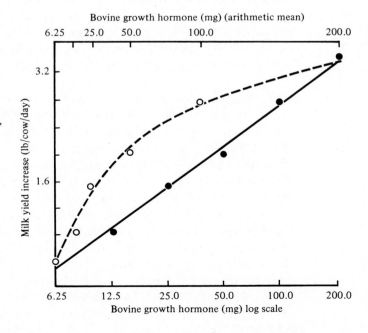

Fig. 8.22. Effect of increasing doses of bovine growth hormone on the milk yield of cows. Upper curve represents doses plotted on an arithmetic scale, and lower curve doses plotted on a logarithmic scale. (From J. B. Hutton. *J. Endocr.* **16**, 115 (1957).)

common in women and dogs, and not uncommon in cats. As discussed in Chapter 6, there are three factors that increase the risk of developing breast cancer: an early age at puberty, a late age at first pregnancy, and a late age at menopause. Women who have had a child before the age of 17 have about one-third the breast cancer risk of mothers whose first birth is delayed until 35 years or later; indeed, a first child after the age of 35 increases the risk over that of nulliparous women (see Fig. 6.9). There is a very low incidence of breast cancer in Japanese women in Japan, but the incidence increases when Japanese women emigrate to Western countries, perhaps because of a change in their reproductive life histories. Another possible factor is the presence of hormones or carcinogens in certain foods; a recent survey in the USA indicated a statistical connection between the consumption of dairy products and deaths from breast cancer. Statistical correlations, however, do not necessarily imply causality.

Human lactation

Unlike the situation in other mammals, human lactation is strongly influenced by social and cultural factors, and in recent times women have often chosen not to breast-feed. In Western cultures for the last 3000 years, some mothers have always opted out of breast-feeding. This was only possible when an alternative source of infant food was available: for a long time the only acceptable alternative was for the child to be breast-fed by a foster mother. The wet-nurse was either a slave or hired servant and so the choice of opting out was essentially confined to women in the wealthier classes.

Early in the present century, however, advances in technology and hygiene made it safe to replace the wet-nurse with the feeding-bottle and to rear the child on cow's milk or cow's milk derivatives. The choice of giving up breast-feeding was then no longer a privilege of the rich. Babies appeared to thrive on ruminant milk, and by the 1960s in some Western countries fewer than one in four babies were being breast-fed by the time the mother left the maternity hospital. The situation was regarded by many as a major step forward in the emancipation of women. Others were less enthusiastic. It became evident that the chemical and immunological characteristics of milk were highly species specific, so that attempts to 'humanize' ruminant milks were something of a biochemical nonsense. Furthermore, behavioural studies in animals and man revealed that the mother–infant bonding associated with breast-feeding had beneficial effects on the subsequent behavioural development of the young. Suspicions therefore grew that bottle-feeding might well be producing children who were nutritionally, immunologically and psychologically underprivileged. Serious alarm about the consequences of bottle-feeding, however, was not triggered until mothers in developing countries, subjected to subtle commercial pressures, followed the fashions of their more affluent sisters and gave up breast-feeding only to discover that, in their environment of inadequate hygiene, the feeding-bottle and formula milk became lethal.

As the potential dangers of bottle-feeding became manifest the question was asked: 'Why do human mothers readily opt out of an act that is instinctive for all other mammalian mothers?' It was Mavis Gunther who first stressed some 25 years ago that it is mimicry in man and higher primates rather than instinct that informs the mother how to feed her offspring – thus the abandonment of breast-feeding, once introduced, tends to be self-perpetuating. Over the centuries mothers have given numerous reasons for not wishing to breast-feed. Examples from the 17th century include: breast-feeding is unfashionable, it may injure one's health, it may ruin the figure, it is noisome to one's clothes, it interferes with gadding about and visiting the theatre. Substitute 'work' for 'gadding about' and the list could have been written today. Only in *Homo sapiens* does the mammary gland have a function additional to its capacity to secrete milk. As already noted, the oestrogen-sensitive stroma of the human mammary gland grows rapidly during puberty to form the protruding breast, whose initial functions are to act as a sex attractant and as a sensitive receptor organ for sexual stimuli (see Book 6, Chapter 1, and Book 8, Chapter 1). When the time arrives for the mammary gland to serve its nutritive role of secreting milk for the nurture of the young, association of the sexual functions of the gland may arouse conflicting feelings in the mind of the mother, and she may attach to the act all her sexual inhibitions and taboos. As the Newtons and the Jelliffes observe 'the sensuous nature of breast-feeding has seldom been recognized in Western society' and that it is 'psychohormonally pleasurable is not surprising. The survival of mankind has depended upon the satisfaction gained by two voluntary acts in the reproductive cycle – coitus and breast-feeding.'

In several developed countries breast-feeding is on the increase again. Certain women's organizations, such as La Leche League, and The Nursing Mothers' Association of Australia, are playing a major role in educational programmes and are promoting feeding on demand rather than on a strictly time-scheduled basis. Peter Hartmann and his colleagues in Perth, Western Australia, have described how in 1972 less than 50 per cent of mothers leaving maternity hospitals were breast-feeding their babies, but by 1980 80 per cent of mothers were doing so and a high percentage of these mothers continued to breast-feed for six months or longer. In the highest socio-economic group, 88 per cent of infants were being breast-fed at six months of age, and 27 per cent were still being breast-fed at 12 months of age. In the lowest socio-economic group the comparable figures were 46 and 13 per cent. Hartmann concludes that 'these findings show that a large and historically important change has occurred in Australia over the last nine years. Affluent well-nourished mothers, en masse, have begun to breast-feed successfully their babies over long periods of time'. It may be noted that the average daily milk yield of these Australian mothers feeding on demand during the first six months was over 1.1 kg, which is higher than the widely accepted WHO/FAO norm of 0.7–0.9 kg per day. The potential for milk production was even greater, since mothers breast-feeding twins were producing 2–3 kg milk per day (Fig. 8.23), indicating that in mothers nursing single babies it is the baby's requirement for nourishment rather than the woman's capacity to produce milk that determines the amount secreted.

The lactational performance of these Australian mothers is not exceptional for, as Hartmann notes, enlightened mothers in other countries are also able to breast-feed and maintain adequate growth of their infants over long periods of time. Motivation clearly plays a major role in ensuring successful human lactation. Not only can inhibitory attitudes and cultural and social fashions be overcome but so also can more basic factors such as stress and fear that may inhibit lactation; mothers have successfully breast-fed their babies under conditions of great stress and food shortage in wartime in occupied countries and in internment camps.

The study of lactational physiology has hitherto been aimed at a greater understanding of the process in the economically important dairy animals. Although many of the observations had potential clinical significance, they aroused little interest; in most medical textbooks lactation received scant attention, being tacked on, almost as an afterthought, to the chapter on reproduction. A change now seems underway. There is a growing appreciation of the biological importance of lactation, of the uniqueness of milk, which is specifically tailored to the nutritional and immunological requirements of the young of each species, and of the important role that suckling may have on mother–infant bonding and on birth spacing. Lactation is indeed a masterly reproductive strategy.

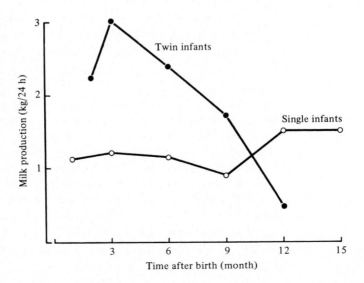

Fig. 8.23. Milk production in women breast-feeding twins for 12 months, and single infants for 15 months. (By courtesy of Dr Peter Hartmann.)

Twin infants

Single infants

Milk production (kg/24 h)

Time after birth (month)

Suggested further reading

A history of infant feeding. I. G. Wickes. *Archives of Disease in Childhood*, **28**, 151–8, 232–40, 332–40, 416–22, 495–502 (1953).

Overview of the mammary gland. A. T. Cowie. *Journal of Investigative Dermatology*, **63**, 2–9 (1974).

The significance of lactation in the evolution of mammals. C. M. Pond. *Evolution*, **31**, 177–99 (1977).

Suckling. E. M. Blass and M. H. Teicher. *Science*, **210**, 15–22 (1980).

Long-term effects of perinatal exposure to hormones on normal and neoplastic mammary growth in rodents: a review. T. Mori, H. Nagasawa and H. A. Bern. *Journal of Environmental Pathology and Toxicology*, **3**, 191–205 (1980).

Preparation and culture of mammary gland explants. R. Dils and I. A. Forsyth. *Methods in Enzymology*, **72**, 724–42 (1981).

Breastfeeding and reproduction in women in Western Australia – a review. P. E. Hartmann, J. K. Kulski, S. Rattigan, C. G. Prosser and L. Saint. *Birth and the Family Journal*, **8**, 215–26 (1981).

Milk ejection in a marsupial, *Macropus agilis*. D. W. Lincoln and M. B. Renfree. *Nature*, **289**, 504–6 (1981).

Mammary gland growth and milk ejection in the agile wallaby, *Macropus agilis*, displaying concurrent asynchronous lactation. D. W. Lincoln and M. B. Renfree. *Journal of Reproduction and Fertility*, **63**, 193–203 (1981).

Breast-milk production in Australian women. S. Rattigan, A. V. Ghisalberti and P. E. Hartmann. *British Journal of Nutrition*, **45**, 243–9 (1981).

The composition of milk. R. Jenness. In *Lactation*, vol. 3. Ed. B. L. Larson and V. R. Smith, pp. 3–107. Academic Press; New York and London (1974).

Comparative Aspects of Lactation. M. Peaker. Zoological Society of London and Academic Press; London (1977).

Human Milk in the Modern World. D. B. Jelliffe and E. F. P. Jelliffe. Oxford University Press; Oxford, New York, Toronto (1978).

Physiology of Mammary Glands. A. Yokoyama, H. Mizuno and H. Nagasawa. Japan Scientific Societies Press, Tokyo; University Park Press, Baltimore (1978).

Machine Milking. C. C. Thiel and F. H. Dodd. National Institute for Research in Dairying, Reading; Hannah Research Institute, Ayr (1979).

Hormonal Control of Lactation. A. T. Cowie, I. A. Forsyth and I. C. Hart. Springer-Verlag; Berlin, Heidelberg, New York (1980).

Growth and differentiation of mammary glands. Isabel A. Forsyth. In *Oxford Reviews of Reproductive Biology*, vol. 4, pp. 47–85. Ed. C. A. Finn. Clarendon Press; Oxford.

The biological basis for the contraceptive effects of breast feeding. R. V. Short. In *Advances in International Maternal and Child Health*, vol. 3, pp. 27–39, ed. D. B. Jelliffe and E. P. Jelliffe. Oxford University Press (1983).

Physiological Strategies in Lactation. M. Peaker, R. G. Vernon and C. H. Knight. Symposia of the Zoological Society of London, No. 51. Academic Press; London, New York (1983).

INDEX